Acknowledgements

This book began life under the supervision of the late Peter Lipton, and it is with the deepest regret that I note that its final execution has been all the poorer for his loss. It is in thanks for his many years of guidance – and perhaps more importantly, his enduring philosophical example – that I dedicate this work to him.

I would also like to extend special thanks to James Ladyman, who has provided both inspiration and foil for my ideas – in his published work, and in a particularly illuminating doctoral examination. Many thanks also to Tim Lewens for pastoral as well as intellectual support; and of course to Bas van Fraassen, not only for his wealth of philosophical argument, but also for his gracious encouragement of my no doubt deeply misguided efforts.

Parts of this work have been delivered at seminars at Bristol University, the University of Cambridge and at the National University of Singapore; my thanks to all those present for their questions and comments. Some of the arguments in Chapter 3 are based upon Dicken, P. (2006) 'Can the Constructive Empiricist Be a Nominalist? Quasi-Truth, Commitment and Consistency', *Studies in History and Philosophy of Science* **37**, pp. 191–209 and Dicken, P. (2009) 'On the Syntax and Semantics of Observability: A Reply to Muller and van Fraassen', *Analysis* **69**, pp. 38–42; a much condensed version of the overall argument can be found in Dicken, P. (2009) 'Constructive Empiricism and the Vices of Voluntarism', *International Journal of Philosophical Studies* **17**, pp. 189–201. Parts of Chapter 4 are based upon Dicken, P. (2007) 'Constructive Empiricism and the Metaphysics of Modality', *British Journal for the Philosophy of Science* **58**, pp. 605–612.

I have been privileged to work with some extraordinarily fine minds at the University of Cambridge, both in the Faculty of Philosophy and in the Department of History and Philosophy of Science: I would like to thank Steve John, Mark Sprevak and Florian Steinberger for especially constructive feedback on my project; thanks also to Yoon Choi, Axel Gelfert, Arash Pessian and Nick Tosh for their incredulity and collegiality in equal measures. Special thanks to Fred D'Agostino for detailed comments on the manuscript as a whole. I would also like to thank Craig Bourne and Marina Frasca-Spada for beginning my philosophical education all those years ago. Finally, my thanks to Matt Kramer, a *sui generis* philosophical entity; and to the Master and Fellows of Churchill College, where this work was completed as a junior research fellow.

On a more personal note, my deepest thanks to my parents who have continued to support me in a countless number of ways (many, but not all of them, financial). And to Katrina Gulliver, who makes it all worthwhile.

Paul Dicken
Churchill College, Cambridge

1
Arguments Concerning Constructive Empiricism

1.1 The aim of science

Should scientists believe everything they say? Ought they to believe the claims of their mature scientific theories, and in the existence of the various microscopic exotica that are now said to populate the unobservable reaches of reality? Or would a more modest attitude towards scientific inquiry be preferable? On first reflection, there is certainly a strong intuition that our current scientific theories are (at least approximately) true: after all, contemporary scientific practice is enormously successful in terms of both the prediction and the manipulation of empirical phenomena, and – so the thought goes – this fact would simply be *miraculous* if it was not the case that our current scientific theories provide us with more or less accurate descriptions of the way the world is. Whatever else one may think about it, science *works*; therefore we should believe that it is true.

But the inference from scientific success to scientific truth is a problematic one. There is certainly no *logical* connection between the two, since it is clearly possible for a predictively successful scientific theory to be false. Take an example from the history of science, say, Newtonian mechanics. At its inception, this was an undoubtedly successful scientific theory that correctly predicted a range of exciting new phenomena – indeed, the theory is so successful that we still rely upon it today when attempting to blast various bits of debris into orbit. But by our contemporary standards the theory is also straightforwardly false, since it deals with a range of discredited concepts such as absolute space, absolute time and absolute velocity that we have now come to reject. In fact, the history of science seems replete with examples of predictively successful theories now considered to have been conclusively shown

1

to be false. And this raises a second, more troubling, intuition about the truth of science: if all of these successful theories of the past have turned out to be wrong, why should we suppose that our own situation is any better? Philosophers of science in Newton's day were presumably as impressed by the predictive success of their scientific theories as we are today with ours; but if their inference from success to truth led them astray, on what grounds should we be any more confident?

These two conflicting intuitions – from the success of current scientific practice and the failure of past scientific practice, respectively – frame the contemporary philosophical debate over whether or not we should believe the pronouncements of our contemporary scientific theories. Scientific realism is the view that we should believe them: that our intuitions concerning the success of our contemporary scientific theories do give us good reasons to think that they are true; and that the historical track record of our scientific theorising is misconstrued if one takes it to undermine our current epistemological position (our past scientific theories may have been false, they argue, but since they have also all been *improvements* upon their predecessors, our current situation is actually rather healthy). Constructive empiricism – a position first put forward by, and forever closely associated with, the iconoclastic philosopher of science Bas van Fraassen – offers a more guarded assessment, for the constructive empiricist believes *some* of the claims made by our scientific theories, while simply suspending judgement about the rest. More specifically, constructive empiricism is the view that 'science aims to give us theories that are empirically adequate; and acceptance of a theory involves as belief only that it is empirically adequate' (van Fraassen, 1980: 12). A theory is empirically adequate provided it gets it right about the observable phenomena: the constructive empiricist therefore only takes our contemporary scientific theories to be in the business of delivering the truth about the medium-sized dry goods with which we are all familiar; and while these scientific theories may well deal indispensably with the microscopic exotica that prompted our initial misgivings, it is not part of the philosophical understanding of scientific practice that we must believe that such entities and processes really exist.

This contrast can be made more explicit by the natural temptation to understand scientific realism – and by extension, the various positions throughout the philosophy of science that seek to offer alternative views regarding the aim of science – in terms of three distinct philosophical theses: a metaphysical claim about the nature of the world investigated by our scientific practices; a semantic claim about the language of the scientific theories we produce in the process of this investigation; and an epistemological claim concerning to what degree the latter can be said to give us knowledge about the former.[1] The metaphysical component of scientific realism states that there is a mind-independent physical world which scientific inquiry seeks to describe; this would be contrasted with, say, a radical social constructivism that argues that the nature of the physical world is somehow *determined* by our scientific practices, rather than merely *discovered* by them.[2] The semantic component of scientific realism states that the language of our scientific theories is to be taken more or less at face value, and that the theoretical terms of our scientific theories that purport to refer to various unobservable entities and processes are indeed to be understood as referential terms. The contrast here would be with the logical positivist's briefly held view that since theoretical terms acquire their meaning through their relationship with more readily testable 'observational' terms, putative reference to unobservable entities is in fact reducible to complex conditionals involving only observable states of affairs;[3] or the instrumentalist view that statements involving theoretical terms are not in fact assertoric at all, and that therefore questions of reference simply never arise.[4]

The epistemological component of scientific realism states that our scientific theories are (at least approximately) true, and that therefore the claims of our accepted scientific theories are (on the whole) a source of knowledge about the unobservable world; this view is usually endorsed on the strength of the sort of explanatory considerations with which we began. Since the constructive empiricist agrees with the scientific realist in terms of both the metaphysical view that there is a mind-independent physical world which it is the purpose of our scientific practices to investigate and the semantic view that our scientific theories at least purport to refer to unobservable entities and processes, it is natural to understand constructive empiricism simply in terms of the denial of the epistemological component of scientific realism – the constructive empiricist, more sanguine at the possibility of a successful yet false scientific theory, is merely *sceptical* (or perhaps better, merely *pessimistic*) about the results of our scientific practices, although he agrees with the scientific realist about the details of the practices over which they differ in enthusiasm.

Yet while the above taxonomy is not exactly *wrong*, it does threaten to obscure what is most important and most interesting about the constructive empiricist's position. For while the constructive empiricist does differ from the scientific realist in only believing our contemporary scientific theories to be empirically adequate rather than true, to characterise constructive empiricism as a kind of sceptical scientific realism fails to do justice to the broader epistemological framework of which this view of scientific practice is a part. Constructive empiricism – as articulated and defended by van Fraassen – is at heart not so much a view about scientific success as a view about scientific *rationality*: it is a view about what it is to make an epistemic judgement (i.e., not with whether or not we should believe our scientific theories to be true, but with what it means to hold a particular belief about the nature of a scientific theory in the first place); it is a view about the obligations of being a rational scientific agent (i.e., not with whether or not scientific truth is the best explanation of scientific success, but with whether or not our best explanations are rationally compelling); and it is a view about how best to understand the very philosophical debate concerning the aim of science with which we are concerned (more specifically, with whether or not the disagreement between scientific realists and constructive empiricists is a substantive philosophical matter, or merely an expression of taste). As a result, much of the standard debate over constructive empiricism – and in particular, the opposition between the scientific realist's optimistic assessment of scientific practice and the constructive empiricist's apparently more curmudgeonly attitude – simply flies wide of the mark. A sceptical scientific realist owes us an account of how he can come to hold different beliefs about the pedigree of our scientific theories, resist otherwise compelling lines of argument over the success of our scientific practice and accommodate our intuitive convictions concerning the nature of the scientific enterprise within his more parsimonious perspective; by contrast, the constructive empiricist needs to defend his views on what it is to hold a belief about a scientific theory, motivate his account of what makes a particular method of inference compelling and advocate his alternative understanding of what a philosophical view about the aim of science seeks to provide. And while this may well be far from being an *easier* task, it is at least an importantly *different* one.

To defend constructive empiricism then is to defend a particular re-conceptualisation of the epistemic framework in which we should approach our initial question about scientific belief. And it is with this broader epistemological issue that this book is concerned: not with whether or not the scientific realist or the constructive empiricist is right, or with whether or not we ought to believe our scientific theories to be true, but rather with what the appropriate epistemic framework might be for conducting such a debate in the first place. For while I agree with van Fraassen that an empiricist conception of the aim of science must be part and parcel of an empiricist conception of the background epistemology of the philosophy of science, I disagree that his own proposed framework – 'epistemic voluntarism', a permissive conception of our inferential practices that demands little more for the rationality of a set of beliefs than its logical consistency – provides the most hospitable environment for constructive empiricism. Indeed, I argue that the combination of constructive empiricism and epistemic voluntarism is as it stands *untenable*, on the grounds that there exists a systemic difficulty in establishing the internal coherence of constructive empiricism that the minimal epistemological resources of epistemic voluntarism are particularly unable to resolve. There is however an alternative epistemic framework, embryonic in van Fraassen's discussion, that promises a much more satisfactory approach to issues of this nature – an epistemic framework that focuses not so much on the different standards of rationality that may be in play in the philosophy of science, but on the different kinds of *attitude* that one can hold towards a proposition or theory. I argue that this approach to the epistemological dimension of the scientific realism debate – one that takes seriously the distinction between acceptance and belief – not only provides the best framework for articulating and defending constructive empiricism but also motivates an attractive empiricist strategy with respect to a range of other philosophical topics.

I begin in this chapter with a more detailed exposition of constructive empiricism, and in particular, why the articulation and defence of an empiricist conception of the aim of science must *ipso facto* be the articulation and defence of a distinctively empiricist epistemology: in the rest of this chapter I shall illustrate the general point about the need to understand constructive empiricism within the context of a broader epistemological framework by first discussing some of the more prominent criticisms that have been levelled against the position, and then how these criticisms are transformed within the context of van Fraassen's epistemic voluntarism. Chapter 2 will be devoted to the exposition and evaluation of the various arguments that can be adduced in support of van Fraassen's epistemic voluntarism, and Chapter 3 will develop the claim that any adequate defence of constructive empiricism must in fact abandon this framework altogether (or at least, that any adequate defence of constructive empiricism must proceed *independently* of van Fraassen's epistemic voluntarism, thereby rendering such a framework superfluous). Only with these preliminaries in place can we begin the task of constructing an improved epistemic framework for the empiricist understanding of science, developed and discussed in Chapter 4.

1.2   The constructive empiricist and his critics

1.2.1   A taxonomy of objections

Constructive empiricism then is the view that the aim of science is empirical adequacy rather than truth, and that to accept a successful scientific theory need not commit us to the belief that such a theory is true. Immediately, this raises two important questions: one concerning the notion of empirical adequacy, and one concerning the notion of acceptance. A scientific theory is empirically adequate just in case it is accurate with respect to the observable phenomena (whereas a scientific theory is true just in case it is accurate with respect to both the observable *and* the unobservable phenomena); but just what is it for a phenomenon – that is, an entity, process or anything else that our scientific theories purport to accommodate – to be observable? For van Fraassen at least, this is an empirical distinction, to be determined by our best scientific theories of the behaviour of light and the physiology of the human eye (van Fraassen, 1980: 17). Yet even leaving the vagaries of such a criterion aside, one may well wonder what relevance such anthropocentric concerns have for the philosophy of science: just what is it about the naked eye that can lead us to privilege what we can observe over what we can (merely) detect with the aid of our scientific instruments? In short, if the constructive empiricist is going to maintain that the aim of science is empirical adequacy rather than truth, then he must defend a principled distinction between the observable and the unobservable phenomena with which our scientific theories are concerned; and we may well worry about the grounds for such a distinction.

Suppose however that such a defence was forthcoming; the constructive empiricist would then maintain that when we accept a successful scientific theory, our confidence in that theory need only extend as far as that which we can observe (this being specified by his concept of empirical adequacy). But if this is the case, then the acceptance of a scientific theory is something distinct from the belief in that theory, since *ex hypothesi* the constructive empiricist need only believe a subset of that which he accepts. So in addition to a principled distinction between observable and unobservable phenomena, the constructive empiricist must also defend a principled distinction between the attitudes of acceptance and belief; and we may similarly worry about the grounds for this distinction too.

In this latter case – regarding the purported distinction between acceptance and belief – the problem can be put as follows: on the one hand, since the constructive empiricist concedes that the unobservable

content of our scientific theories is indeed indispensable to their predictive and manipulative success, to accept a scientific theory must
therefore involve a substantial epistemological commitment to that
theory sufficient for the constructive empiricist to in some way accommodate those parts of the theory that he does not believe (or else face a
hopelessly impoverished account of scientific practice that essentially
rejects everything our scientific theories say regarding unobservable
phenomena); yet on the other hand, since the constructive empiricist
maintains that the aim of science is limited to empirical adequacy, the
acceptance of a scientific theory must therefore not involve *so much* epistemological commitment that it simply collapses into the putatively
contrasting attitude of belief (in which case constructive empiricism
would amount to a merely verbal alternative to scientific realism). The
proper response to this objection must however be delayed until a later
section (Chapter 4), since it is in teasing out the nuances of the distinction between acceptance and belief that we are able to construct
a more general epistemic framework for the articulation and defence
of constructive empiricism. In particular, as we shall see, it is only by
attending to the distinction between acceptance and belief that the constructive empiricist can maintain his other crucial distinction between
the observable and the unobservable; and it is this that ultimately justifies abandoning van Fraassen's epistemic voluntarism in favour of an
alternative epistemic framework.

With respect to the first distinction between observable and unobservable phenomena, we can further clarify the dialectical situation by
distinguishing between two different types of objection that the critic of
constructive empiricism may raise: we can worry about whether or not
the distinction between observable and unobservable phenomena is *any
good*, that is, with whether or not it draws a philosophically relevant distinction with respect to our understanding of scientific practice; and we
can worry about whether or not such a distinction even *makes sense*, that
is, with whether or not it is even logically consistent for the constructive empiricist to draw such a distinction in the first place. This latter
objection was first formulated by Alan Musgrave (1985), who notes that
since the constructive empiricist must rely upon our scientific theories
in order to determine the extension of what counts as an observable
phenomena (specifically our best scientific theories of the behaviour of
light and physiology of the human eye), yet wishes to use this result in
order to draw a distinction that determines which parts of these very
scientific theories are to be believed, he is at risk of defending a philosophical position that undermines its own cogency. The issue of the

internal coherence of constructive empiricism is a particularly tough one, and must again be delayed until a later section (Chapter 3); as it will transpire, this is the crucial issue that will ultimately show the inadequacies of van Fraassen's epistemic voluntarism and motivate the development of an alternative epistemic framework based upon the various attitudes in play within the philosophy of science, as opposed to the various standards of rationality.

All of which leaves us, for the time being, with the issue of relevance. The problem here is that if the constructive empiricist is going to maintain that he need only believe what his scientific theories say regarding observable phenomena in order to make sense of scientific practice, then he appears to be committed to an epistemological distinction between the observable and the unobservable consequences of his scientific theories – the idea being that if we are only to believe what our scientific theories say about the observable, then we must possess *greater justification* for believing what our scientific theories say about the observable than we do for what they say about the unobservable. The difficulty then lies in the fact that there seems to be very little of epistemological relevance to be found between what our theories tell us about very big things and what they tell us about very small things.

Such reasoning however brings us inextricably to the issue of the appropriate epistemic framework in which to conduct our debate. For to criticise the constructive empiricist's distinction between observable and unobservable phenomena on the grounds that our beliefs about the two are equally justified is to presuppose an epistemological context in which one should apportion equivalent degrees of belief in response to equivalent degrees of evidence – an epistemic framework governed by rules of inference that one is obliged to follow in all those circumstances in which they can be applied, an epistemic framework where one *should* believe whatever one has reasonable grounds to believe. And while this is certainly the epistemic framework of the scientific realist – who is prepared to infer the truth of his scientific theories on the basis of their predictive success as an instance of the general policy of inferring the truth of one's best explanations – this is not the epistemic framework of the constructive empiricist since, as we shall see, van Fraassen's epistemic voluntarism is *precisely* the denial that rationality is ever a matter of obligation. Thus, in short, even if his critic is correct that our beliefs about unobservables are just as justified as our beliefs about observables, it does not follow that the constructive empiricist must abandon his distinction between the two, since it is not part of the constructive empiricist's understanding of scientific rationality that one must always draw the same sorts of conclusion in the face of the same sorts of evidence.

But this is all moving too quickly; in the remainder of this section I shall consider some of the arguments pertaining to the relative justification of our beliefs about observable and unobservable phenomena, and the prospects of replying to them on their own terms. My survey will be far from complete: I will not be considering, for example, those issues that concern the inherent vagueness of the distinction between observables and unobservables; nor will I be considering the explicitly anthropocentric bias of the constructive empiricist's position (see, e.g., Maxwell, 1962). In part, this is because many of these issues have been discussed in great detail in a number of other sources.[5] More importantly though, this is because my aim in this chapter is to consider those objections that most helpfully illuminate our broader concerns regarding the appropriate epistemological framework in which to pursue the scientific realism debate. I will be considering then those objections to the distinction between observable and unobservable phenomena that claim that our beliefs about the two are *equally justified*, and those objections that claim that the constructive empiricist is following an *inferentially unprincipled* methodology in drawing his distinction. Both objections therefore implicitly assume the sort of rule-based epistemology that constructive empiricism is an explicit attempt to challenge. Having outlined these challenges, I shall then sketch van Fraassen's alternative epistemic framework in more detail, and show how this transforms these various concerns regarding the relevance of the constructive empiricist's distinction. The rest of this book is devoted to a sustained examination of this so-called new epistemology.

1.2.2 Observation and detection

Many philosophers have criticised constructive empiricism on the grounds that its central distinction between observable and unobservable phenomena is not sufficiently well-motivated. Ian Hacking (1985), for example, has attempted to undermine the division, and with it the constructive empiricist's contention that one can accept a successful scientific theory without thereby believing the claims it makes concerning unobservable phenomena, by advancing several arguments for the comparable reliability of both 'direct' and 'mediated' observation. He maintains that it is implausible to suppose that beliefs about unobservable phenomena (acquired via the use of complex scientific instruments, such as the electron microscope) are somehow less justified than beliefs about observable phenomena (acquired via the naked eye) since the former must be 'mediated' or 'interpreted' in a way that the latter need not: he argues that the eye itself can be considered as a kind of sophisticated instrument, as our best scientific theories of human physiology attest; that one must equally learn to use both kinds of instrument before their results can be considered reliable; and that for the experienced microscopist, such usage is as direct and unmediated as normal vision is. As Hacking puts it, 'one needs a theory to make a microscope. You do not need a theory to use one' (1985: 137).

Moreover, Hacking argues that the persistent agreement between various different instrumentally aided acts of observation gives us good evidence for their individual reliability. He gives the example of observing a microscopic object that we ourselves have created through an independent process; in such cases:

> I know that what I see through the microscope is veridical because we *made* [the object in question] to be just that way ... moreover, we can check the results with any kind of microscope, using any of a dozen unrelated physical processes to produce an image. Can we entertain the possibility that, all the same, this is some gigantic coincidence?
>
> (1985: 146–147)

Thus Hacking concludes that there are no compelling grounds to consider the scientific beliefs we acquire regarding unobservable phenomena to be less reliable, justified or warranted than the scientific beliefs we acquire regarding observable phenomena; indeed, since we can acquire these latter beliefs from a diverse range of sources that all register an impressive level of agreement, we actually have good reasons to consider our scientific beliefs regarding unobservable phenomena to be *at least* as reliable, justified or warranted than anything that we acquire through the naked eye. The distinction that the constructive empiricist wishes to draw between observable and unobservable phenomena is therefore epistemologically irrelevant; and if the distinction between the observable and the unobservable fails to hold any water, then the constructive empiricist's view that in order to make sense of scientific practice we need only believe our successful scientific theories to be empirically adequate – and that one can thereby accept a successful scientific theory without thereby believing it to be true – is equally undermined.[6]

By contrast, Paul Churchland (1985) has rejected the epistemological relevance of the distinction between observable and unobservable phenomena, not because he believes that our scientific claims about the latter are just as well justified as our scientific claims about the former, but because he is *equally sceptical* about both. He argues that

> since our observational concepts are just as theory-laden as any other, and since the integrity of those concepts is just as contingent on the integrity of the theories that embed them, our observational ontology is rendered *exactly as dubious* as our non-observational ontology.
>
> (1985: 36 [original emphasis])

Churchland then proceeds to adduce support for this contention by noting historical examples of mistaken ontologies in both the observable and the unobservable domain: he notes that 'we have had occasion to banish phlogiston, caloric, and the luminiferous ether from our ontologies – but we have also had occasion to banish witches and the starry sphere that turns about us daily', the latter being 'as "observable" as you please and … widely "observed" on a daily basis' (1985: 36).

Before proceeding, it should be noted immediately that such considerations are far from conclusive. Churchland, for example, simply overstates his case: for while it may indeed be the case that both our observable and unobservable ontologies have had their shares of winners and losers, the most that his argument can establish is the *qualitative* claim that both our observable and unobservable concepts are theory-laden; but in order for his objection to have any force, he must in fact establish the stronger, *quantitative* claim that our observable and unobservable concepts are *equally* theory-laden. But simply adducing examples of discredited observable phenomena to accompany examples of discredited unobservable phenomena is too coarse-grained an approach to achieve this result: what Churchland wishes to establish is that our observable concepts are *just as theory-laden* as our other concepts are; but all he manages to establish is the much weaker conclusion that our observable concepts are *theory-laden, just as* our other concepts are.

Churchland's argument therefore, while perhaps questioning the degree of justification enjoyed by our observational ontology, does not rule out the possibility that our observational ontology is still *better justified* than our other ontologies – and so even by the scientific realist's own epistemological standards, fails to undermine the constructive empiricist's purported distinction. And similarly in response to Hacking, neither does the immediacy of microscopic observation guarantee its epistemic reliability. As with Churchland's misgivings, the argument again merely establishes a qualitative relationship between our acts of unaided observation and instrumental detection, a phenomenological similarity with respect to the apparent transparency of both methods. But in this case, while it may make little sense to distinguish between the various degrees in which observation and detection can be immediately given, the problem simply lies in the fact that our phenomenological *impressions* of the reliability of a method do not guarantee its *actual* reliability; and thus, even by the scientific realist's own epistemological standards, it does not follow that the constructive empiricist's distinction is undermined, since it may be the case that unaided observation is the only *veridically* immediate method (this point is developed in more detail in van Fraassen, 2008: 105–109).

The strongest line of argument with which we are left then concerns the apparent continuity of our unaided observations and instrumentally aided detections – that our microscopic observations confirm the evidence of our own eyes, and that our various distinct detection techniques all register an impressive level of agreement. Yet we should be careful in how we interpret these results. In particular, we should not underestimate the difficulty and ingenuity involved in producing an image or representation through a scientific instrument, even when dealing with something as apparently straightforward as a microscope (a nice antidote to Hacking's account is developed in detail in van Fraassen 2008: 91–113). Scientific instruments must be manipulated in order to produce anything that we could reasonably take to be a representation, and often these representations tell us more about the scientific instrument in question. At the very least, we must concede that our scientific instruments *create new phenomena* just as much as they (purportedly) represent existing phenomena: new phenomena such as the fact that this particular arrangement of lens produces this particular image for this particular observer (in these particular circumstances), an observable phenomena that must be accommodated within our empirically adequate theories just as much as any other. The real issue then is not whether or not our scientific instruments furnish us immediately and directly with images, but whether or not these images are better thought of as windows onto the unobservable or artifices of the instrument itself. As van Fraassen puts it,

> The point I am making is that the microscope *need not* be thought of as a window, but is *most certainly* an engine creating new optical phenomena. It is accurate to say of what we see in the microscope that we are 'seeing an image' (like 'seeing a reflection', 'seeing a rainbow'), and that the image could be *either* a copy of a real thing not visible

to the naked eye or a mere public hallucination. I suggest that it is moreover accurate and in fact more illuminating to keep neutrality in this respect and just think of the images themselves as a public hallucination.

(2008: 108–109)

More importantly, the fact that any act of instrumentally aided observation will depend upon our ability to manipulate and fine-tune the scientific instrument in question (as Hacking say, one *does* need a theory to make a microscope . . .) undermines any conclusion that the scientific realist might draw on the basis of the impressive level of agreement that exists between all of our myriad detection techniques. As van Fraassen argues (1985: 298), the persistent agreement of diverse instrumentally aided observations need not be considered all that surprising, or in need of any profound explanation, once one takes into account how all of our scientific instruments rely upon one another for calibration. Hacking's argument would be more compelling if every single act of instrumentally aided observation recorded the same results. But this is manifestly not the case: getting a scientific instrument to produce an image or representation is hard work, and the only grounds upon which we would be willing to consider a particular piece of apparatus as so doing would be for it to produce the same sorts of result as those other pieces of equipment that we already deemed as capable of detecting unobservable phenomena. But if this is the case, then Hacking's argument loses much of its force, since if our primary reasons for even considering something as a scientific instrument is that it agrees with all of our other scientific instruments, we cannot then *appeal* to the fact that our scientific instruments all agree as an argument for their individual reliability!

All of which invites two further, more general, comments. The first is to note that the disagreement over our instrumental reliability is just a microcosm of the disagreement between realists and empiricists in general: whereas the realist argues that the predictive and manipulative success of all of our various scientific theories gives us good reasons to suppose that our scientific reasoning *in general* is successful at getting at the truth, the empiricist replies that the predictive and manipulative success of all of our various scientific theories is hardly surprising since any *unsuccessful* theory would never have been included in the list (van Fraassen, 1980: 40) – instrumental agreement does not indicate reliability, since it is only those pieces of equipment that agree that we classify as being a scientific instrument; contemporary predictive success does not indicate truth, since it is only those theories that are predictively

successful that survive to be contemporary. Thus just as with our instruments, so too with our theories: the general phenomena cannot entail
anything interesting about their specific instances, since our interpretation of the general phenomena will in fact presuppose what it is about
their particular instances that we wish to show – Are all of our theories predictively successful because they are true, or because we select
for predictive success? Do all of our instruments agree because they are
reliable, or because we select for agreement?

We may however still wish to ask about individual successful theories
and individual reliable instruments – Why is *this* theory predictively
successful? Why do the images produced by *this* instrument aid my
manipulation of the observable phenomena? – And this brings us to the
second point. For what underlies Hacking's argument, as it does Churchland's argument, is a very specific epistemological assumption denied by
the constructive empiricist. For Churchland, the assumption is that one
should hold the same epistemological attitude towards those claims that
share the same epistemological *justification* (although as we have seen, it
is unclear that what our scientific theories say regarding observable and
unobservable phenomena do share equal justification). For Hacking, the
assumption is that the veridicality of our instruments – and the verac-
ity of our theories – is the best explanation of their utility (although
we have also seen reasons to doubt this too), and that inference to the
best explanation is a *rule* of rationality that the constructive empiricist is
perverse to disregard. It is these assumptions that really characterise the
scientific realism debate, and which van Fraassen's epistemology seeks
to reject.

1.2.3   Extra terrestrials and extra-terrestrials

Both Hacking and Churchland mount cases to the effect that what our
scientific theories say regarding unobservable phenomena is epistemically on a par with what our scientific theories say regarding observable
phenomena – on the grounds that our beliefs about both are just as
immediately given, and just as vulnerable to revision, as the other. Yet
as we have seen, not only are there reasons to doubt these claims, there
are even greater reasons to doubt the conclusions drawn from them: in
short, that the justificatory equivalence of what our scientific theories
say regarding observable and unobservable phenomena is only philosophically significant on the assumption that our epistemological attitudes are completely determined (and strictly limited) by the inferences
we make upon our evidence. The situation is even more apparent in the
case of our scientific instruments: Hacking's considerations concerning

the persistent agreement of diverse instrumental detections are only as strong as the intuition that the best explanation for our instrumental success is that they provide faithful windows onto the unobservable world; and not only have we seen reasons to doubt *this* claim, it is also clear that it could only have argumentative force on the assumption that we must *always* infer the truth of our best explanations.

In outlining the potentially problematic assumptions driving his critics' arguments, we can already begin to see the sort of epistemological picture favoured by the constructive empiricist – a view whereby our epistemological attitudes are determined as much by our *values* as they are by our evidence, and where an inferential rule may be *permissible* but never obligatory. The contrast is made even clearer when we consider those objections to the constructive empiricist's distinction that target not the supposed justificatory inequality of what our scientific theories say regarding observable and unobservable phenomena, but the alleged inferential double-standards that lead to this supposed inequality.

Churchland's article is again particularly instructive in this respect: for in support of his universal pessimism regarding the representational success of our scientific theories, he offers two general lines of attack at what he takes to be the inferential double-standards at the heart of van Fraassen's position. The first concerns the constructive empiricist's apparently arbitrary attitude towards counterfactuals (Churchland, 1985: 39–40; see also Psillos, 1999: 193–200). In order to draw a distinction between observable and unobservable phenomena, the constructive empiricist must appeal to what we *could* have observed in the appropriate circumstances – rather than simply to what we *have* observed – or else risk drawing an absurdly narrow and hopelessly impoverished distinction. The constructive empiricist must therefore endorse certain counterfactual possibilities while rejecting others. For example, the distant moon of an alien planet is presumably an observable phenomenon if anything is, even if it is in fact so distant that it will never actually be observed; the constructive empiricist must therefore endorse the claim that *if there were* a suitably placed human observer (looking out the window of some intergalactic spacecraft, say), *he would* be able to observe the moon. Similarly, a long-extinct but sufficiently intimidating dinosaur is no less an observable phenomenon simply for having existed before mankind evolved; and so the constructive empiricist must also endorse the claim that *if there had been* a prehistoric observer, *he would have* been able to see it; and so on and so forth. By contrast, sub-atomic particles are supposed to be unobservable *tout court*; the constructive empiricist must therefore roundly reject the counterfactual possibility of a microscopic observer meeting such phenomena face-to-face. But there seems to be no principled way to make such a distinction, and no epistemic justification as to why one counterfactual possibility is closer to us in logical space than another (is the possible world where we invent a radical new method of intergalactic transport really more similar to the actual world than the one where we invent some radical new method of shrinking?). In sum, the constructive empiricist's distinction between the observable and the unobservable seems quite arbitrary.[7]

The second line of thought invites us to consider those whose observational ontologies are very different from ours – the notorious aliens with electron microscopes for eyes. In this case, the constructive empiricist faces a simple dilemma: while on the one hand, it seems that he should concede that such instrumentally endowed creatures can directly observe various microscopic phenomena (or else have his distinction succumb to his epistemic chauvinism), on the other hand, it seems that the constructive empiricist cannot concede that such creatures can directly observe microscopic phenomena – the causal mechanism whereby the aliens observe microscopic phenomena is exactly the same as the causal mechanism whereby we observe microscopic phenomena (the only salient difference between the two cases is the epistemologically innocuous one that whereas the aliens have their electron microscopes permanently fixed to their faces, we can leave ours in the laboratory), therefore if microscopic phenomena are observable for the aliens, then they must be observable for us too (Churchland, 1985: 43–44). Thus just as the constructive empiricist endorses some counterfactual inferences while rejecting other – apparently equivalent – counterfactual inferences, so too does he allow that certain causal mechanisms mediate direct observations while denying that other – apparently equivalent – causal mechanisms do not. A shrinking machine is no more or less outlandish than an intergalactic spaceship, and an electron microscope that happens to be glued to one's face is still an electron microscope; the constructive empiricist's distinction between observable and unobservable phenomena therefore seems to rest upon nothing more than straightforward epistemological dishonesty.

Yet just as with those objections that pertained to the alleged justificatory equivalence of the constructive empiricist's observable and unobservable ontologies, so too can we point to weaknesses in these arguments concerning the constructive empiricist's alleged inferential inconstancy. In response to the first objection – that the constructive empiricist should believe what his scientific theories say regarding unobservable phenomena, since this requires no greater leap of counterfactual faith than it takes to believe what they say regarding observable phenomena (or conversely, that the constructive empiricist should withhold his belief in anything other than what his scientific theories say regarding the merely *observed* phenomena) – we can note that there are in fact perfectly good reasons for supposing a distant moon or extinct dinosaur to be observable-in-principle that do not translate into good reasons for supposing sub-atomic particles to be observable-in-principle. Recall that for the constructive empiricist, the distinction between observable and unobservable phenomena is an empirical distinction, to be determined by our best scientific theories regarding the behaviour of light and the physiology of the human eye. These scientific theories have a lot to tell us about the various parameters that determine whether or not something can be observed: the size of the object to be observed; the frequency of the electromagnetic radiation scattered by, or emitted from, the object in question; the resolving power of the observer's eye; and so on and so forth. Crucially though, none of these theories contain any parameters for where exactly in time and space the observational event is supposed to take place. Thus we have good reasons to suppose that our distant moon is observable, precisely because our best scientific theories tell us that an object that large can be observed by the human eye, and that simply being onboard some fancy spacecraft a long way away from Earth will make absolutely no difference to the emission and absorption of electromagnetic radiation. By contrast, however, we have equally good reasons to suppose that sub-atomic particles *cannot* be considered as observable-in-principle, precisely because our best scientific theories tell us that the sort of enormous physical transformation entailed by the shrinking scenario envisaged above is *exactly* the sort of variation that will have a significant effect upon our sensory modalities. Or to put the point another way: the constructive empiricist's distinction between observable and unobservable phenomena is to be determined by our best scientific theories *relative* to a particular epistemic community (i.e., with respect to the physiology of the *human* eye); thus even if it transpired that a sufficiently altered observer could encounter sub-atomic particles directly (although given what we know about the behaviour of light, this seems unlikely), it would be irrelevant for our present purposes since such an observer could no longer tell us anything interesting about *our* epistemic community.

This issue concerning the appropriate scope of our epistemic community is also important with respect to Churchland's second argument.

This is the objection that, since he would grant that aliens with micro-scopes for eyes can directly observe microscopic phenomena, yet deny that human beings with detachable microscopes can directly observe such phenomena, the constructive empiricist is forced into the uncom-fortable conclusion that aliens and humans 'must embrace *different* epistemic attitudes towards the microworld, even though their causal connections to the world and their continuing experience of it be iden-tical' (Churchland, 1985: 44 [original emphasis]). But as van Fraassen (1985: 256–258) points out, there is an equivocation in this argument. If we were to encounter such a race of instrumentally augmented aliens, two possibilities would then present themselves. The first is that, in the spirit of intergalactic goodwill, we would welcome these individu-als into our epistemic community – that we would treat them, for all intents and purposes, as 'one of us'. Were this to happen, our notion of observability would be extended accordingly: since some members of our epistemic community would now be able to make direct observa-tions of microscopic phenomena, such phenomena would thereby fall within the scope of what counts as observable-for-us. Of course, such an eventuality may well require a substantial division of epistemic labour, with our peculiar-looking neighbours carrying most of the authority of what we (the community) can and cannot see; but the important point to note is that in such a situation, the constructive empiricist would not be forced to assign different epistemic status to otherwise causally indistinguishable individuals.

The second possibility is that, in the grip of a debilitating xenopho-bia, we refuse to accommodate these instrumentally augmented aliens within our epistemic community. In this case, our notion of observabil-ity would remain the same. Yet in refusing to consider such individuals as members of our epistemic community, the constructive empiricist again avoids any charge of epistemic inconstancy. The basic point is that if we do not consider these aliens as part of our epistemic community, it simply begs the question to suppose that their 'causal connections to the world and their continuing experience of it' *are* identical to ours. From an epistemological point of view, we have no reasons to treat these instrumentally augmented aliens any differently from the more famil-iar – and less animate – electron microscopes that we encounter in the laboratory.

Churchland's objection rests upon conflating these two possibilities. He invites us to imagine a situation where we assume that these instru-mentally augmented aliens possess the same causal connections and continuing experiences as we do, while at the same time refusing to

count them as members of our epistemic community. But to even countenance such a scenario is to beg the question in favour of Churchland's
realism – it presupposes a particular view about the *unobservable mech-*
anisms supposedly common to both microscopic detection and extra-
terrestrial observation that the constructive empiricist simply denies
that we are in a position to know. For Churchland, we have good reasons to believe the deliverances of our scientific instruments regardless
of whether or not we count such instrumentation as part of our epistemic community, since for Churchland there are rationally compelling
inferences to make concerning the predictive and manipulative success
of instrumental detection and extra-terrestrial observation, respectively.
But for the constructive empiricist, such inferences are not rationally
compelling, and the only circumstances in which we would be forced to
take our instrumentally augmented neighbours at their word would *ipso*
facto be circumstances in which we would be forced to reconstrue the
borders of our epistemic community accordingly.

Thus again, neither argument concerning the distinction between
observable and unobservable phenomena need cause the constructive
empiricist undue consternation, even taken on their own terms: the
supposedly arbitrary endorsement of certain outlandish counterfactual
possibilities but not others is in fact grounded upon what our best scientific theories tell us are relevant considerations with respect to observability; and the alleged double-standards regarding otherwise equivalent
causal mechanisms can only be considered as such if one already
favours scientific realism over constructive empiricism. Most importantly though, and again as with those earlier arguments pertaining to
the justificatory equivalence of the constructive empiricist's observable
and unobservable ontologies, these objections also presuppose a specific conception of the background epistemological framework – not
just that evidentially equivalent hypotheses should be treated equally
(i.e., that there is no room for pragmatic or evaluative considerations
in forming one's doxastic attitudes), but moreover that our inferential practices are governed by certain rules that admit of no exception.
Thus similar counterfactual possibilities must be judged similarly likely
or unlikely (although we have of course seen reasons to doubt that
the counterfactual possibilities in question are sufficiently similar), and
equivalent causal mechanisms should be considered equivalently veridi-
cal or misleading (although we have of course seen reasons to doubt
that the mechanisms in question can be considered equivalent without
begging the question at hand). The exceptionless application of obligatory inferential rules to nothing more or less than our evidence – this

is the epistemological framework that underlies these criticisms of constructive empiricism, and which the constructive empiricist explicitly rejects.

1.3   The new epistemology

1.3.1   The English and Prussian models of rationality

So far I have only discussed those elements of the so-called traditional epistemological framework that underlie many of the scientific realist's objections to constructive empiricism, and which van Fraassen's philosophy of science therefore attempts to deny: the idea that rationality is a rule-governed practice whereby certain inferential practices demand our universal assent; and the idea that the process of belief revision is completely exhausted by the systematic application of these inferential methods to our bodies of evidence. Yet to note which elements of this familiar epistemological picture that the constructive empiricist rejects is not in itself very informative – nor for that matter, very convincing – without a positive characterisation of those elements of his alternative epistemological framework that he does endorse.

   In Chapter 2, I offer a more detailed appraisal of van Fraassen's epistemic voluntarism, and in particular the specific arguments that he advances for what transpires to be an extremely minimalist conception of rationality. But for our present purposes, we can begin both to motivate and to illuminate this epistemic perspective by locating it within the context of some broader and more familiar epistemological issues. The first is the venerable problem of scepticism, of how it is that we can have any knowledge of the external world. There are, roughly speaking, two different approaches we can take to this problem (van Fraassen, 1989: 170–171). The first is to concede that our everyday beliefs do stand in need of justification in the face of the sceptic's challenge, and consequently to attempt to find some secure footing upon which we can build our edifice of knowledge. Thus foundationalists attempt to isolate those indubitable ideas which are beyond even the sceptical challenge, and to show how everything else we take ourselves to know can be secured upon this basis. A recent manifestation of this strategy would be the idea of a 'sense datum' – again, a foundational experience that can admit of no doubt – and the project of attempting to 'construct' our more complex experiences out of these meagre materials. The second strategy, by contrast, rejects the sceptical challenge as posing an unrealistic and implausible demand. The idea here is that we already find ourselves out and about in the external world guided by a set of beliefs that we otherwise have no reason to question; after all, we all have to start somewhere, and so the weight of the sceptical challenge is seen to lie not upon us to provide *even more* justification for what we reasonably take to be well-founded beliefs, but upon the sceptic to provide us with specific arguments for rejecting specific beliefs.

The contrast between these two approaches helps to illustrate what van Fraassen has in mind with his alternative epistemological framework. The first response to the sceptical challenge is akin to the traditional epistemology that van Fraassen rejects: in attempting to answer the sceptic directly, those who favour the first strategy are thereby endorsing the idea that what one should believe is (merely) what one can provide sufficient reasons for believing; similarly, in criticising the constructive empiricist's distinction between observable and unobservable phenomena on the grounds of their alleged epistemological equivalence, the scientific realist is endorsing the idea that what one should believe is (exactly) what one can provide sufficient reasons to believe. In both cases, the background epistemological assumption is that our doxastic practices are completely determined by certain inferential rules – we are not to believe anything that has not been generated by the appropriate inferential rule, and we are to believe everything that is generated by the appropriate inferential rule. Belief revision is rendered a systematic, even mechanical, process.

Conversely, the second response to the sceptical challenge is akin to the new epistemological framework that van Fraassen advocates: by refusing to engage with the sceptic directly, those who favour the second strategy are endorsing the idea that what one is justified in believing is more than simply what one can defend in the face of the sceptical challenge; and similarly, in rejecting the criticism that our observable and unobservable ontologies are equally justified, the constructive empiricist is endorsing the idea that one need not believe everything that can be inferred from the data. In both cases, the background epistemological assumption is that our doxastic practices are as much an issue of pragmatics and value as they are of rules and evidence – in van Fraassen's memorable phrase, 'what it is rational to believe includes anything that one is not rationally compelled to disbelieve... *Rationality is only bridled irrationality*' (1989: 171–172 [original emphasis]).

The second, and more specific, epistemological issue that can help us to locate van Fraassen's particular perspective concerns the generation of hypotheses. This issue is especially important for the philosophy of science, but even in everyday contexts we must inevitably form opinions about the world that go beyond our evidence – whether these are assumptions about the uniformity of nature and of how things will evolve in the future, or assumptions about the unobservable reaches of reality that we take to underlie what we do in fact encounter. And again, there are roughly speaking two different approaches that one can take to this issue (van Fraassen, 1989: 172–173): the first is the conservative view that one should only allow those hypotheses that are positively supported by our current evidence; the second is the more liberal option of allowing any hypothesis that is not explicitly ruled out by our current evidence. The first option again relates to the kind of rules-based epistemology that was shown to underlie the scientific realist's objections to constructive empiricism: those who favour the first option of hypothesis generation presuppose that there are certain canons of inferential practice which, when applied to our existing body of evidence, conclusively determine what it is and what it is not rational for us to believe (strict Bayesianism, which tells us precisely how we are to update our subjective probabilities upon the acquisition of new evidence, is perhaps the clearest example of this kind of approach); similarly, when the scientific realist complains that the constructive empiricist is making one sort of inference in one case, while making a different sort of inference in another (as in his allegedly arbitrary conclusions concerning different types of counterfactual possibility), he is presupposing that there is a single canon of inferential practice that the constructive empiricist is thus epistemologically iniquitous for only endorsing as and when it pleases him.

By contrast, the second option relates to van Fraassen's more permissive conception of rationality. In allowing all those hypotheses that have not been explicitly ruled out, those who favour this second option are conceding along with William James that our doxastic practices are intimately related to our doxastic values: we are faced with the conflicting goals of wishing to acquire truth while avoiding error, the first of which can be satisfied by simply believing *everything* (which would therefore maximise our true beliefs), the second of which can be satisfied by simply believing *nothing* (which would therefore minimise our false beliefs); the trick of course is to find a happy medium, and since there can be no algorithm for *that*, whichever way one seeks to balance these conflicting goals will ultimately come down to a matter of personal preference (James, 1948). Similarly, when the constructive empiricist rejects the claim that he must always draw the same sorts of counterfactual conclusion from the same sorts of evidence, or that he must always infer the truth of his best explanations, he is advancing the claim that what we *should* believe (or equivalently, what it is *rational* to believe) cannot be completely dictated by the disinterested application of a set of rules – to argue about the appropriate attitude that we should take towards the various elements of our scientific theories is itself to argue about what we take the aim of science to be; and what we take the aim of science to be is itself an expression of our epistemological values, of our desire for truth and our aversion of error, which themselves will determine when we are prepared to make an inference, and when we are not.

These two themes – that we are justified in our beliefs so long as we have no specific reason to doubt them, and that whatever inferences we wish to draw from our beliefs will be determined as much by our epistemological values as it will be by some abstract dictate of reason – conform to what van Fraassen refers to as the Prussian and English concepts of law (van Fraassen, 1989: 171). Rationality is of course a matter of what one is and is not allowed to do (epistemically speaking), and in the legal sphere there are two contrasting ways to make these specifi-cations. The Prussian concept is a bottom-up approach: it focuses upon what one is *allowed* to do, lists those actions that are to be deemed legal and then specifies that 'everything is forbidden which is not explicitly permitted.' By contrast, the English concept is a top-down approach: it focuses upon what one is *forbidden* to do, lists those actions that are to be deemed a crime and then specifies that 'everything [is] permitted that is not explicitly forbidden' (van Fraassen, 1989: 171). And so it is with the traditional and voluntarist conceptions of rationality. For the scien-tific realist, there are certain beliefs that one is allowed to hold (relative to a particular set of data), and there are certain inferences that one is allowed to make; and anything else (such as the constructive empiricist's doxastic dilettantism) is therefore deemed irrational. For the construc-tive empiricist, there are certain beliefs that one is forbidden to hold (relative to a particular set of data), and there are certain inferences that one is forbidden to make; and anything else (counterfactual inequal-ity, refusing to infer the truth of one's best explanations) is all that one could hope to mean by 'rational'.

1.3.2 Relativism, scepticism and voluntarism

According to van Fraassen's so-called voluntarist epistemology, rational-ity is to be considered a matter of permission rather than obligation, where one is rationally entitled to believe anything that one is not ratio-nally compelled to disbelieve (this characterisation can also be found in van Fraassen (2000: 277; 2002: 92, 97)), or to approach the same point from another angle, where an agent can be considered rational in holding a particular combination of beliefs just in case that combination does not sabotage its own possibility of vindication (van Fraassen, 1985: 248; 1989: 157). What this all boils down to is that once one has shown that one's set of beliefs meet the minimal standards of logical consistency and probabilistic coherence (the minimal criteria for avoiding epistemic self-sabotage), there is simply no further work for a substantive epistemology to do.[8]

However, we still need to consider how exactly this framework allows the constructive empiricist to answer – or rather, finesse – those objections raised by the scientific realist concerning the distinction between observable and unobservable phenomena, or more specifically, how this framework allows the constructive empiricist to answer his critics without thereby undermining the epistemological project altogether. We can put the problem as follows. Any epistemological framework that we adopt will ultimately be concerned with two fundamental issues: the status of our current beliefs, and the method whereby we revise these beliefs. Recalling the analogies drawn in the previous section with the problems of scepticism and hypothesis generation, we can see that van Fraassen's epistemic voluntarism is committed to the claims that there are no independent justifications available for our current beliefs (we cannot help but to start somewhere, and the sceptic is therefore philosophically perverse to demand the justifications that he does), and that there are no independent justifications available for the ampliative inferences we make upon our evidence (if one desires truth at all costs then one should make as many ampliative inferences as one can; if one seeks to avoid error at all costs then one shouldn't make any ampliative inferences at all). It is these two strands that allow him to resist the scientific realist's conclusion: just because one can give an argument as to why one should treat what our scientific theories say regarding observable and unobservable phenomena as epistemically on a par, it does not follow that the constructive empiricist is *irrational* not to do so; and just because one can give an argument as to why the veridicality of our scientific instruments is the best explanation for their reliability, it does not follow that the constructive empiricist is *irrational* not to draw such a conclusion. But if there are no independent justifications available for our current beliefs, and no independent justifications available for how we are to extrapolate from our beliefs, then our epistemic lives seem doomed to either a debilitating scepticism (if we suppose that the notion of justification is something to which we should aspire) or a radical relativism (if we suppose that the notion of justification is something we can happily do without). Either conclusion is unacceptable, and it is important to note how the epistemic voluntarist attempts to navigate between the two.

With respect to the first horn, one must bear in mind that a lack of justification for our current beliefs, and a lack of justification for our methods of ampliative inference, will only entail scepticism if we also suppose that it is irrational to hold a belief (or make an inference) that is not justified. This is certainly the sceptic's assumption, and is an implicit premise in any sceptical argument. Yet it is an assumption that many deny, not just the epistemic voluntarist. A useful comparison here is with the strict Bayesian whom we encountered above. According to Bayesianism, rationality consists in the mechanical updating of our subjective probabilities in the face of new evidence, a process that can be given a precise mathematical formulation. Yet it is no part of the Bayesian machinery – and thus on this view, no part of the epistemological project *per se* – as to the origin of the beliefs that are to be updated; that is to say, one can assign any prior probability that one likes to a proposition, provided one is prepared to update that probability in the manner specified. Thus, just as with the epistemic voluntarist, Bayesians deny that there is any independent justifications available for our current beliefs; yet they also deny that their position succumbs to a debilitating scepticism, since even if we cannot be completely justified in our beliefs, we can at least be completely justified in our assessments of how much more or less likely our beliefs are, conditional on any new piece of evidence. The epistemic voluntarist goes one step further than the strict Bayesian in this respect however. Bayesian conditionalisation is a non-ampliative process – in effect, it is simply a probabilistic constraint upon the distribution of an agent's subjective probabilities at any particular moment of time. Thus the Bayesian agrees with the epistemic voluntarist that there are no independent justifications available for our methods of ampliative inference, and in this case also agrees with the sceptic that it is irrational to use an unjustified method of ampliative inference – the conclusion drawn by the Bayesian is thus that one should abandon all methods of ampliative inference altogether and rely entirely upon those tools furnished by the (non-ampliative) probability calculus. But this of course is to advocate the sort of rules-based epistemology that the epistemic voluntarist is also at pains to deny, contending instead that the way we revise our beliefs is as much a matter of value as it is of rule. Thus just as he denies that it is irrational to hold a belief that is not justified, so too does the epistemic voluntarist deny that it is irrational to make an ampliative inference that is not

justified. Scepticism is therefore avoided, since on the epistemic voluntarist's conception of rationality, there is nothing irrational in the
otherwise unjustified practices with which the sceptic takes issue.

Such considerations however threaten to impale the epistemic voluntarist on the equally unappealing horn of relativism: if it is rational to
hold a belief without any independent justification, and it is rational
to perform any sort of ampliative inference without independent justification, then it may look as if any set of beliefs is just as good as any
other – we lose the capacity to criticise others, and must simply accept
that what is rational for me may not be rational for you. But there are
two points to bear in mind here. The first is that, even upon such a minimal conception of rationality, there still remain the basic constraints of
logical consistency and probabilistic coherence: it is rational to believe
anything that one is not rationally compelled to disbelieve, and a set
of beliefs that guarantee that they cannot all be true together is exactly
the sort of self-refuting combination that one is rationally compelled to
disbelieve. The second point to bear in mind is that rationality is not the
same as truth: two different agents can be rational in holding conflicting beliefs, even though at least one of them must be wrong, since for
the epistemic voluntarist one can be rational in holding a false belief.
And this of course takes us right back to how we initially introduced
constructive empiricism – to accept a scientific theory involves as belief
only that it is empirically adequate, and the issue of whether or not the
theory is *true* (as opposed to a predictively successful, yet false, theory
such as Newtonian mechanics) is simply irrelevant to the philosophical
understanding of scientific practice.

My comments here are of course far from complete, and as we shall
see in Chapter 2, the twin dangers of scepticism and relativism will
return to haunt van Fraassen's advocation of epistemic voluntarism:
two of the central arguments that van Fraassen advances in favour of
his minimal epistemological framework are that the sceptical challenge
renders a more substantive epistemology untenable, and that an epistemological framework that takes seriously the role of our epistemic values
provides a more satisfactory framework for certain meta-philosophical
debates on how to proceed in – for example – the philosophy of science.
Yet in invoking the sceptical considerations necessary for undermining
a traditional rules-based epistemology, van Fraassen also threatens to
undermine his more parsimonious alternative; and in reducing much
of our meta-philosophical disagreements to expressions of taste, van
Fraassen thereby threatens to render such debates irresolvable. It therefore remains an extremely delicate balancing act to both resist the

scientific realist by denying that there exists an independent justification for either our current beliefs or methods of inference and simultaneously resist the sceptic and/or relativist by denying that it is irrational thereby to hold these beliefs and make such inferences.

1.4 Summary

Constructive empiricism – as it has been formulated and developed by its originator, Bas van Fraassen – does not make any epistemological claims about the nature of science: it is a view regarding the *aim* of science, and as such it is entirely neutral with respect to the epistemic attitudes that we should hold towards the consequences of our accepted scientific theories. To be sure, the constructive empiricist is committed to the claim that one does not *need* to believe that our accepted scientific theories are true in order to make sense of scientific practice, since the belief that they are merely empirically adequate is argued to suffice. But claims about what we need to believe do not determine our epistemic policy: as Ladyman et al. (1997: 318–319) point out, one could agree that the aim of science is mere empirical adequacy, agree that we need not believe our accepted scientific theories to be more than empirically adequate in order to accommodate contemporary scientific practice, yet still maintain as a matter of epistemically well-placed fact that our accepted scientific theories are *true*; and conversely, one could deny that the aim of science is mere empirical adequacy, deny that we need only believe our accepted scientific theories to be empirically adequate in order to accommodate contemporary scientific practice, yet still maintain a healthy degree of scepticism about the enterprise as a whole.

More to the point, it is also clear that constructive empiricism – as it has been articulated and defended by van Fraassen – *cannot* make any epistemological claims regarding the nature of science. This follows from the broader epistemological project within which it is to be situated. According to van Fraassen's so-called voluntarist epistemology, rationality is to be considered as a matter of permission rather than obligation, where one is rationally entitled to believe anything that one is not rationally compelled to disbelieve. What this all boils down to is that once one has shown that one's set of beliefs meet the minimal criteria of logical consistency and probabilistic coherence, there simply is no further work for a substantive epistemology to do. Consequently, constructive empiricism cannot be construed as an epistemological claim regarding the nature of science – for example, about which of the claims of our accepted scientific theories we are justified in believing – since within this voluntarist framework, substantive issues of justification and warrant are simply moot.

It is precisely this component of van Fraassen's philosophy of science that is the main focus of this book. For while it is clear that a view regarding the aim of science will be perfectly neutral with respect to any view regarding the success of science, it seems equally clear that any view regarding the aim of science will be intimately related to a specific view concerning the philosophy of science. As we have seen in the examples of Churchland and Hacking, to defend scientific realism is *ipso facto* to defend a particular conception of scientific rationality – a conception according to which we are rational to believe only that which we have good reasons to believe, and where the process of belief revision is exhausted by the disinterested application of an unquestionable canon of rules. Consequently, to defend constructive empiricism is also to defend a particular conception of scientific rationality, of how we are to even frame the scientific realism debate – and in van Fraassen's view, this is to defend a conception of rationality whereby we are rational to believe whatever we do not have good reasons not to believe, and where our methods of inference are to be understood as permissible rather than obligatory. Yet while I agree with the premise of van Fraassen's argument, I disagree with the conclusion: to defend a particular view regarding the aim of science is to defend a particular view of the philosophy of science; but I disagree that the distinction between a 'traditional' rules-based conception of rationality and a 'voluntarist' values-based epistemology is the most auspicious manner in which to pursue this issue.

In Chapter 2, I shall discuss van Fraassen's epistemic voluntarism in more detail, and in particular examine the various arguments he advances for this minimalist conception of rationality. While my assessment of these arguments will be fairly negative, the main critical work will be undertaken in Chapter 3, where I argue that the combination of constructive empiricism and epistemic voluntarism is as it stands untenable: in order to defend the internal coherence of his view regarding the aim of science, the constructive empiricist must appeal to more epistemological resources than are provided by epistemic voluntarism. In Chapter 4, I sketch just such a framework, one based not upon the various standards of rationality that one can bring to bear upon the philosophy of science, but upon the various attitudes that one can hold towards a proposition or theory – a distinction already implicit in van Fraassen's presentation of constructive empiricism, although as yet insufficiently developed. Thus while much of the following discussion will be devoted to criticising his specific views, the guiding premise of this work is that *van Fraassen is essentially right*, and that one cannot make progress in the philosophy of science without attending to the background epistemological presuppositions that one brings to bear upon the debate. This thought is interesting enough to warrant a sustained examination in its own right, and with luck, it may even help to further the constructive empiricist cause.

2
Epistemic Voluntarism: Rationality, Inference and Empiricism

2.1 Motivating voluntarism

The work of Chapter 1 was to bring to the fore the importance of van Fraassen's broader epistemological framework in understanding his articulation and defence of constructive empiricism. To put the point as succinctly as possible, since constructive empiricism is merely a view about the aim of science – rather than a substantive thesis about its epistemological limits – it is consistent with being a constructive empiricist to recognise any possible position with respect to the success of science in giving us knowledge about the physical world: that it aims for empirical adequacy does not in itself tell us to what extent scientific inquiry realises, falls short of or even exceeds this goal. Consequently, any argument that attempts to discredit constructive empiricism on the grounds that we have good reasons to believe that our contemporary scientific theories tell us a great deal beyond the empirically adequate simply fails to engage with the position.

That constructive empiricism must be understood as this arguably more modest epistemological proposal (as opposed to a rigorously selective scepticism that argues that our scientific theories produce truths only insofar as they refer to observable phenomena) follows from van Fraassen's views on the nature of epistemology in general. According to van Fraassen, rationality is a matter of permission rather than obligation, such that one can only really criticise a course of cognitive action if it is guaranteed to sabotage its own chances of success. This allows us to generalise the point made above: it can be no criticism of constructive empiricism that our scientific inquiry outperforms mere empirical adequacy (and note that we have yet even to see a satisfactory argument to this effect), since the constructive empiricist is not committed to any view concerning the success of science; and the reason why the constructive empiricist abjures any discussion concerning the success of science is that, from a voluntarist perspective, seeking to maximise truth is no better and no worse than seeking to minimise error. To put the point even more simply, if the only standards of rationality for the epistemic voluntarist is to avoid self-contradiction, then there is no sense in which forming true beliefs about the unobservable realm is any more of a cognitive achievement than refusing to speculate about the matter altogether, and thus no sense in which the various inferential practices employed by the scientific realist to this effect could possibly demand our universal assent. Scientific realists have one set of epistemic preferences, insofar as they pursue a wide range of beliefs formed on the basis of various rationally compelling ampliative inferences; they seek to offer ever deeper levels of explanation for the observable phenomena which they encounter, and often proceed through the postulation of ever more unobservable structure. Constructive empiricists have another set of epistemic preferences, insofar as they pursue a more limited range of beliefs formed on the basis of various rationally permissible ampliative inferences; they reject certain explanatory questions as requiring an answer (although of course one could still choose to answer them), and reject certain explanatory strategies as being even able to provide such answers in the first place. Crucially though, there are no objective criteria for adjudicating between these conflicting preferences, and any argument one way or the other will serve only to express the disputant's own epistemological idiosyncrasies.

It is therefore important to ask what reasons there are for adopting epistemic voluntarism. Although the position has developed gradually over the course of his wide-ranging and voluminous writings, it is possible to distinguish three distinct lines of thought that together constitute van Fraassen's motivation for his more permissive conception of rationality. The first is the most technical, and concerns the probabilistic coherence of an agent's beliefs. For while van Fraassen argues that logical consistency and probabilistic coherence pretty much exhaust the constraints upon an agent's rationality, he also maintains that probabilistic coherence is a more substantial constraint than it is usually given credit. More specifically, while most philosophers would accept that an agent's beliefs should be *synchronically* coherent – that for any particular moment of time, an agent's total distribution of credences at that time should satisfy the probability calculus – van Fraassen argues that in addition, an agent's beliefs should also be *diachronically* coherent – that for any particular moment of time, an agent's total distribution of credences at that time will also constrain his total distribution of credences at a later time. The precise respect in which an agent's beliefs at one time will constrain his beliefs at a later time is captured in what van Fraassen calls the Reflection Principle; and the relationship between diachronic coherence and epistemic voluntarism lies in the claim that it is only by adopting the latter that one can understand what holding a belief amounts to, given the additional constraints imposed by the former. This is the *positive* argument for epistemic voluntarism: that the constraints of diachronic coherence force us to understand the making of an epistemic judgement as the undertaking of a *commitment* to certain future courses of cognitive action, as exemplified in the voluntarist notion of pursuing certain epistemic preferences.

The second line of thought is by contrast the least technical, and is in effect an expression of van Fraassen's pessimism towards the prospects of a more substantive epistemology. If the traditional (non-voluntarist) epistemologists are right, then in any particular context, there will be specific rules of inference that any rational agent is obliged to make upon his evidence. But according to van Fraassen, there are no inferential rules or practices that meet this requirement; hence there can be no meaningful question as to what a rational agent is *obliged* to do. This is the context in which van Fraassen advances his frequently misunderstood critique of inference to the best explanation. It is often complained that the constructive empiricist cannot object to the scientific realist's inference to the truth of his successful theories, since the constructive empiricist's inference to the empirical adequacy of a theory is just as precarious. The point of the argument however is not that abductive reasoning can *never* be justified; rather, it is that since abductive reasoning cannot be shown *always* to be justified, it cannot be supposed to be a rationally compelling method of inference. This is the *negative* argument for epistemic voluntarism: that it is left merely to an agent's epistemic preferences the degree to which various ampliative methods of inference are to be employed; and that the only principles that can demand their universal assent are the constraints of logical consistency and probabilistic coherence.

These two lines of thought – the positive argument from diachronic coherence, and the negative argument against ampliative inference – are the basic ingredients of van Fraassen's notion of an *epistemic stance*, the voluntarist alternative to a canon of inferential practice. Together, they also propel the third line of argument motivating epistemic voluntarism, which is that it does better justice to the nature of various metaphilosophical debates that have persisted throughout the discipline.

Essentially, van Fraassen argues that certain broad philosophical positions – his favourite examples are empiricism and materialism – do not make sense construed as the advocation of a dogma, or the possession of a fundamental set of core beliefs. Moreover, disputes between competing philosophical positions cannot be adjudicated in the same way as those disputes that arise *within* a particular position, since there will be little or no shared background assumptions, and no higher court of appeal (this of course is the diagnosis of what goes wrong in the debate between scientific realists and constructive empiricists). Broad philosophical positions like empiricism are better construed as a choice of one's epistemic policy, coupled with a commitment to certain future courses of cognitive action – the negative and positive arguments for epistemic voluntarism, respectively.

In this chapter, I shall assess these various arguments for epistemic voluntarism. My conclusion will be rather negative. In the case of the Reflection Principle, I shall argue that despite van Fraassen's various lines of defence, diachronic coherence leads to absurd results; moreover, that those cases that van Fraassen does offer where diachronic coherence appears to be a virtue are in fact better understood as *misdescribed* instances of the importance of synchronic coherence. More importantly, van Fraassen's principal case against the justification of our ampliative practices is a sceptical one, and here he faces the delicate balancing act anticipated in the previous chapter. On the one hand, he must motivate sufficient scepticism to bring our ampliative inferences into disrepute; yet on the other hand, he must not invoke so much scepticism that he renders all epistemological debate redundant – his minimal conception of rationality notwithstanding. I argue that this enterprise fails, and that any argument strong enough to dissuade us from making ampliative inferences will likewise dissuade us from adopting any epistemic framework whatsoever. This particular worry also finds expression in van Fraassen's notion of an epistemic stance which, while both interesting and fruitful, is currently too permissive in that it leads to a problematic form of relativism.

Of course, rejecting van Fraassen's positive arguments for epistemic voluntarism hardly constitutes a compelling argument *against* the position – that is the work of Chapter 3, where I show that whatever else one concludes about van Fraassen's permissive notion of rationality, it is as it stands a framework fundamentally unacceptable for the advocacy of constructive empiricism. This is an initially surprising claim, especially given the time spent in Chapter 1 demonstrating the role of epistemic voluntarism in diffusing some of the most pressing challenges to constructive empiricism. Nevertheless, I shall show that while epistemic voluntarism may provide the resources to rebut any objection as to the epistemological relevance or desirability of the constructive empiricist's basic position, it is substantially impotent to secure its internal coherence. The argument of this chapter is that, since there are few independently compelling grounds for being an epistemic voluntarist anyway, there may be nothing much lost in drawing this conclusion.

2.2   Diachronic coherence and the principle of reflection

2.2.1   The diachronic Dutch-book

Let us begin with the positive argument for epistemic voluntarism: that independently motivated probabilistic considerations compel us to adopt a principle of diachronic coherence as a constraint upon any rational agent's degrees of belief; and that in order to understand our doxastic practices in the light of this additional constraint, one must take the assertion of one's beliefs to do more than to report an autobiographical fact about one's mental life – it is to undertake a commitment to certain future courses of action in such a way as to inseparably entwine questions of epistemology with issues of pragmatics.

This idea is originally developed in the context of a *diachronic Dutch-book* argument (van Fraassen, 1984)[1] – that unless an agent's distribution of subjective probabilities at any particular moment of time is constrained by his distribution of subjective probabilities at any other moment of time, that agent will be vulnerable to a series of wagers individually acceptable but collectively guaranteed to lose him money. Let H be any hypothesis about which I hold an opinion regarding its likelihood of truth – in van Fraassen's example, the theory of evolution. And let E be the proposition that at some later date, say, tomorrow, I will be fully convinced (i.e., subjective probability = 1) of the truth of hypothesis H. Simple epistemic caution entails that my degree of belief in E – my subjective probability about coming to be certain of the truth of the theory of evolution tomorrow – need not be particularly high. After all, tomorrow may bring counterexamples or the formulation of an even better explanation of the complexity of the natural world; for the sake of argument, let $P(E) = 0.4$. Furthermore, a healthy humility regarding my own cognitive faculties entails that my subjective probability of coming to be certain about the theory of evolution tomorrow, while unbeknownst to me the theory turns out to be in fact false, is not zero; let $P(E \ \& \ \neg H) = 0.2$. Both probabilities seem *prima facie* perfectly reasonable, and indeed, one may even argue that maintaining a healthy degree of scepticism concerning the accuracy of one's own future opinions is a highly desirable epistemic virtue. But from these simple probabilities, van Fraassen argues, it is possible to construct a very straightforward strategy for fleecing me at the track.

The strategy revolves around the following three wagers. The first wager costs me 0.2 units, and will pay out 1 unit in the case that I come to believe that H is true while H is in fact false (i.e., E & ¬H). The second wager costs me 0.3 units, and pays out 0.5 units in the case that I fail to come to believe that H (irrespective of its actual truth-value). The final wager costs me 0.2 units, and pays out 0.5 units in the case that I come to believe that H (again, irrespective of its truth-value). A quick calculation of my expectations for each wager shows that I would consider each wager as perfectly fair.[2] Nevertheless, having taken all three bets together I'm guaranteed to lose money, come what may.

Suppose that I do not come to believe that H; in this case I win the second wager (+0.5 units), but lose the first and third – given that the total cost of the three bets was 0.7 units, I'm out of pocket. But now consider the other possibility: that I *do* come to believe that H. In this case I win the third wager (+0.5 units), and lose the second wager; most importantly however, since *ex hypothesi* I now come to believe that H, I will assume that I have also lost the first wager (or at least, be indifferent to selling that bet on for next to nothing) – so again, given that the total cost of the three bets was 0.7 units, I'm out of pocket. What appeared to be the perfectly rational position of entertaining reasonable doubt about my own future opinions has landed me into probabilistic incoherence (not to mention serious debt at the race track).

The diachronic Dutch-book strategy can be readily generalised, so as to avoid the simplifications made above where my subjective probability in a proposition can be equal to 1, or where we are dealing with propositions whose truth-value will not be decided (van Fraassen, 1984: 241–243). The basic idea is that rather than offering bets on whether or not I will come to believe a proposition with certainty, our cunning bookie offers bets on whether or not I will come to accept as fair a future set of odds on a particular proposition whose truth-value we will be in a position to later ascertain (e.g., whether or not I will come to accept as fair certain odds on a horse winning a race). The strategy then proceeds as before, with one wager paying out if I do come to accept those odds as fair, one wager paying out if I don't come to accept those odds as fair and a final crucial wager paying out if I come to accept those odds as fair *and* the horse loses. The point of course is that should I come to accept these *new* sets of odds, I will be indifferent to selling on this final wager at a value sufficiently different from that on the basis of which I originally accepted the bet (i.e., on the basis of my *old* expectation of the horse winning) such that I am again guaranteed to lose money.

Enter the Reflection Principle. What the above argument is meant to establish is that a probabilistically coherent distribution of beliefs must satisfy both synchronic and diachronic constraints – our beliefs at one moment of time must somehow circumscribe our beliefs at another moment of time, or else we will find ourselves accepting just the sort of wagers presented above. This is what the Reflection Principle provides, and it concerns our degrees of belief *about* our degrees of belief. In essence, it states that my current degree of belief in a proposition A – given the assumption that my degree of belief in that proposition will be a specific value r at a later date – should itself be equal to r; or in other words, for any proposition about which I have an opinion about my own future credence, my current credence in that proposition *conditional on my opinion about my future credence* should be of the same value as that future credence. More formally:

$$P_t(A/P_{t+1}(A) = r) = r$$

where P_t is my subjective probability at a time t, and P_{t+1} is my subjective probability at a later time $t + 1$. In those cases where I am *certain* what my future subjective probability for a proposition will be, my current subjective probability for that proposition should simply match that value. For example, suppose that I am certain that at a later time $t + 1$ I will believe a certain proposition A to a degree of 0.6. By the Reflection Principle, my conditional probability for A at an earlier time t is also equal to 0.6:

$$P_t(A/P_{t+1}(A) = 0.6) = 0.6$$

By the standard definition of conditional probability, this gives us:

$$\frac{P_t(A \& P_{t+1}(A) = 0.6)}{P_t(P_{t+1}(A) = 0.6)} = 0.6$$

However, since I am currently certain about my future credence, $P_t(P_{t+1}(A) = 0.6) = 1$, which by simple substitution means that $P_t(A) = 0.6$.

With respect to the diachronic Dutch-book strategy presented above, the solution lies in the fact that if I had conformed to the Reflection

Principle I never would have had a distribution of subjective probabilities against which such a strategy could be run. In our example, my subjective probability that tomorrow I will come to firmly believe the theory of evolution is equal to 0.4. By the Reflection Principle, this means that my current conditional probability for the theory of evolution – given that my future subjective probability for the theory of evolution is equal to 0.4 – is also equal to 0.4. Moreover, since in the example I am certain about what my future credence will be, my current (non-conditional) subjective probability for the theory of evolution is also equal to 0.4. My subjective probability for coming to believe the theory of evolution even though the theory of evolution is false should therefore be:

$$P_t(E) \times P_t(\neg H) = 0.4 \times 0.6 = 0.24$$

and not 0.2 as stated above. In short, the distribution of subjective probabilities in the diachronic Dutch-book argument is inconsistent with the Reflection Principle; and any agent who respects the Reflection Principle will therefore be immune from the sort of incoherence these arguments generate. The final steps of the argument then proceed rather swiftly. The way to avoid the sort of Dutch-book strategies sketched above is to adopt the Reflection Principle as an additional constraint upon one's doxastic deliberations; but the only way one can justify the Reflection Principle is by conceiving of an epistemic judgement as more than an autobiographical statement, but as something more akin to a speech-act that commits one to a particular course of action. As van Fraassen puts it,

> I conclude that my integrity, *qua* judging agent, requires that, if I am presently asked to express my opinion about whether *A* will come true, on the supposition that I will think it likely tomorrow morning, I must stand by my own cognitive *engagement* as much as I stand by my own expressions of commitment of any sort.... I can no more say that I regard *A* as unlikely on the supposition that tomorrow morning I shall express my high expectation of *A*, than I can today make the statement on the supposition that tomorrow morning I shall promise to bring it about that *A*.
>
> (van Fraassen, 1984: 255)

2.2.2 From Dutch-books to cognitive calibration

The Reflection Principle is therefore initially motivated in response to the possibility of falling foul to a diachronic Dutch-book strategy; in later work however, van Fraassen distances himself from this style of reasoning, on the grounds that it presupposes certain decision-theoretic assumptions – for example, that an agent's distribution of subjective probabilities can be modelled by his betting behaviour – that needlessly complicate the issue (van Fraassen, 1995a: 12). Diachronic Dutch-books are to be considered more as a useful *heuristic* for van Fraassen's views on epistemology rather than a key argument in favour of them. Yet before turning to his more considered reasoning about the Reflection Principle – which involves its relationship to the notion of an agent's beliefs being perfectly calibrated – it is worth exploring a deeper problem with the diachronic Dutch-book strategy that has nothing to do with decision-theoretic assumptions, and which will help to illustrate the general difficulties facing the Reflection Principle.

As my exposition of van Fraassen's argument makes clear, the diachronic Dutch-book strategy trades upon offering the putatively incoherent agent a wager on the basis of his set of beliefs at one time, buying that wager back on the basis of his set of beliefs at a later time and using the difference between these two sets of beliefs as the basis for the incoherence. For example, in the simplified case above, the crucial wager on whether or not I will come to believe the theory of evolution when that theory is in fact false is evaluated not so much on the *truth* of whether or not I falsely come to believe in evolution, but on my *future opinion* as to whether or not I have falsely come to believe the theory – the grounds upon which I accept the wager are different from the grounds upon which I evaluate the wager; or in other words, the wager equivocates over which set of beliefs is in question. The basic worry then is that while the diachronic Dutch-book may demonstrate probabilistic incoherence between a set of beliefs, it doesn't demonstrate probabilistic incoherence between a set of beliefs *that an agent holds at any one time* (Howson, 2000: 135–139). But while we may agree that an agent's set of beliefs *at any one time* should be probabilistically coherent, it is not clear why a rational distribution of credences should seek to avoid the sort of incoherence present in the diachronic Dutch-book strategy: after all, it cannot be a general constraint upon my epistemic behaviour that my beliefs at any one time must be coherent with my beliefs at any other time, as that would be to demand that I never change my mind about what might be true! The point then is that without an independent argument as to why a rational agent should seek to avoid diachronic incoherence, van Fraassen cannot argue for the Reflection Principle on the basis of his diachronic Dutch-books, since the plausibility of the diachronic Dutch-book strategy in fact presupposes the very constraint upon rationality that the Reflection Principle seeks to impose (Christensen, 1991: 242–243).[3]

Diachronic Dutch-book strategies thus provide a doubly contentious motivation for the Reflection Principle (and by extension, epistemic voluntarism in general): not only do they presuppose the decision-theoretic assumptions that van Fraassen notes, they also presuppose the irrationality of certain doxastic states of affairs (i.e., certain temporally extended distributions of subjective probabilities vulnerable to partic-ular betting strategies) that we may in fact be epistemically sanguine about. Let us turn then to van Fraassen's preferred argument for the Reflection Principle, which concerns the notion of how well *calibrated* an agent's beliefs are. Calibration is a measure of agreement between an agent's judgements and the actual frequencies of the events in question: for example, if I believe that every day has a 50 per cent chance of rain, my beliefs will be perfectly calibrated – with respect to a certain interval range – if out of every day in that interval, exactly 50 per cent of them are indeed rainy. In general, an agent's beliefs about the probability of rain will be perfectly calibrated if 'for every r, the proportion of rainy days among those days on which he announces probability r for rain, equals r' (van Fraassen, 1984: 245). Perfect calibration is a difficult feat for an epistemic agent to achieve, yet it clearly seems to be a virtue one should try to attain; conversely, any strategy that ruined *a priori* one's chances for perfect calibration should be rejected as irrational. The con-sidered argument for the Reflection Principle then proceeds by showing how any violation of that principle would jeopardise an agent's chances for perfect calibration; the Reflection Principle is thus justified as a prin-ciple of rationality, and epistemic voluntarism vindicated, again on the grounds that it is the only way to make sense of such a principle.

Keeping to our theme of weather prediction, van Fraassen offers the following examples of a violation of the Reflection Principle that is thereby a violation of calibration. The first (1984: 245–246) is a case where our hapless weather forecaster announces a probability of rain of 0.8, and then announces that for any day for which he announces a probability of rain of 0.8, the probability of rain for that day is in fact 0.7. As can be clearly seen, such a strategy cannot hope to be perfectly calibrated: perfect calibration for the first announcement (exactly 80 per cent of those days on which he makes the first announcement are rainy) precludes perfect calibration for the second announcement (exactly 70 per cent of those same days on which he makes the first announce-ment are rainy), and *vice versa*; in either case therefore, the agent's beliefs as a whole cannot be perfectly calibrated. The second example

(1995a: 13–17) is slightly more straightforward, although essentially the same in structure. Here, our unfortunate weather forecaster announces in the evening the probability of rain for the next day as 0.8, yet knows already that by his own updating procedures he will announce the probability of rain in the morning as 0.2. If such an eventuality is repeated, then it would have to rain tomorrow 80 per cent of the time for his evening prediction to be perfectly calibrated, while only rain tomorrow 20 per cent of the time for his morning prediction to be perfectly calibrated, and again, perfect calibration for the forecaster's beliefs as a whole will be impossible.

In these examples, it seems perfectly clear that our weather forecaster is guilty of some kind of irrationality, and that his distribution of beliefs is in violation of the Reflection Principle – in both cases, the weather forecaster's earlier subjective probability for rain does not match his later subjective probability for rain. What is contentious however is van Fraassen's conclusion that our weather forecaster is irrational *because* of his violation of the Reflection Principle. The worry is easier to see in the second example: if our weather forecaster knows *now* that his own procedures for updating his subjective probabilities will change his prediction from 80 per cent to 20 per cent, then he must know *now* what additional piece of evidence it is that will cause him to revise his prediction (... 'the wind is starting to drop'). But if he knows that piece of evidence *now*, then presumably this will be something that his evening prediction should be taking into account. Our weather forecaster's announcement in the evening for the probability of rain and his announcement in the morning for the probability of rain should therefore be the same, and thus satisfy the Reflection Principle. Crucially though, the reason that these two predictions should match will not be because of the additional constraint of diachronic coherence; it will be for the far more mundane reason that both predictions will be based on *exactly the same evidence* (Engel, 1997). Similar is the case where our weather forecaster makes a prediction about the reliability of his own predictions: if he is in a position to announce that whenever he makes a certain prediction about the likelihood of rain, the actual probability of rain will be something else, then he must be in a position to know what additional factors guarantee this discrepancy between his announced probabilities and the actual probabilities. But if he's in a position to know *this*, then his original announcement should have taken these additional factors into account. His earlier probability of rain should match his later probability of rain, and therefore satisfy the Reflection Principle; but again, the reason that his beliefs satisfy the Reflection Principle is not because diachronic coherence is a necessary constraint upon rationality, but merely as a consequence of both beliefs being based upon exactly the same evidence.

The argument from the diachronic Dutch-book strategy was found wanting since although it challenged the coherence of a set of beliefs, it did not challenge the coherence of a set of beliefs that a rational agent holds at any one time: it compares two distinct sets of beliefs that we have no independently motivated reason to suppose need to be jointly coherent, and complains that their predictions do not match. By contrast, the argument from calibration takes what are in effect two instances of the *same* set of beliefs (since both early and late predictions are based upon exactly the same evidence) and attempts to show that since calibration demands that their predictions match, a rational agent's beliefs must in general meet a constraint of diachronic coherence. On the one hand, we have an illegitimate comparison between what are legitimately considered as two distinct sets of beliefs; and on the other hand, we have a legitimate comparison between what are illegitimately considered as two distinct sets of beliefs. Or to put the point more bluntly: the argument from the diachronic Dutch-book strategy fails because the only reason we might have to suppose that such a temporally extended distribution of subjective probabilities is irrational is if we presuppose the constraint of diachronic coherence such an argument is meant to establish; and the argument from calibration fails since what it demonstrates is not the important of *diachronic* coherence, but rather the importance of *synchronic* coherence, given that we diagnosed the irrationality of the weather forecaster simply in terms of not taking all of his current evidence into account when making his predictions.

2.2.3 Greek sailors, drunk drivers and other epistemic villains

Neither of van Fraassen's arguments therefore establishes the Reflection Principle as a legitimate constraint upon rationality. Moreover, there are a number of damaging counterexamples to the Reflection Principle that must also be taken into account. Consider a situation where an agent knows (for simplicity, is certain) that in the future he will suffer from some kind of impairment to his intellectual faculties, which will result in his coming to believe a proposition that he currently considers to be absurd: he may know, for example, that he has just drunk a glass of Kool-Aid spiked with some psychedelic pharmaceutical which will shortly cause him to believe that he can fly (Christensen, 1991: 234–236); or he may know that he is about to go on a drinking spree, the result of which will be his firm conviction that copious alcohol consumption has no negative effect upon his ability to drive (Maher, 1992: 120–121, 122); or to take the classic example, Ulysses knows that he will shortly hear the Siren's song, and will believe that heading for the rocks is actually quite a good idea (van Fraassen, 1995a: 20). By the Reflection Principle, it seems that in all of these cases the rational agent should believe *now* that he can fly, that he can drive home safely while intoxicated and that heading for the rocks is actually quite a good idea; and this is clearly not the case. Not only then is the Reflection Principle poorly motivated, it also seems to be manifestly false.

Yet as van Fraassen (1995a: 22–26) argues, one must be careful with how exactly one constructs these putative counterexamples. For while on the one hand, a successful counterexample to the Reflection Principle must describe a situation in which the diachronically incoherent sets of beliefs (drink-driving is dangerous; drink-driving is safe) can be coherently attributed *to the same agent*, on the other hand, any successful counterexample to the Reflection Principle must also describe a situation in which the agent's transition from his earlier set of beliefs to his later set of beliefs has an *independently motivated claim to being rational*, regardless of any considerations we may have concerning diachronic coherence. In slightly more concrete terms, some counterexamples to the Reflection Principle are ill-founded because there's no sense in which the epistemic agent in question would recognise these future (and by his current standards, absurd) opinions about flying, drink-driving and setting sail for the Siren's island as *his* future opinions; and some counterexamples to the Reflection Principle are ill-founded because the method by which the epistemic agent in question acquires these future opinions (be that by hallucinogens, alcohol or magical enchantment) is sufficiently irresponsible that the issue of satisfying the Reflection Principle (or indeed, any other principle of rationality) is simply moot. These two responses van Fraassen terms the 'death and disability' defence and the 'integrity' defence, respectively; and he proposes that *any* putative counterexample to the Reflection Principle will be caught on either horn of the dilemma that they present (van Fraassen, 1995a: 25–26).

To take the first horn first, van Fraassen argues quite reasonably that

> The question is not only what foreseen transitions – ways of changing my mind – I, the subject, classify as pathological or reasonable. The question is also what I am willing to classify as future opinions of mine. When I imagine myself at some future time talking in my sleep, or repeating (with every sign of personal conviction) what the

torturer dictates or the hypnotist has planted as post-hypnotic suggestion, am I seeing this as myself expressing my opinions as they are then? I think not.

(1995a: 23)

This is a pretty compelling response in the case of Ulysses, for example, where we might reasonably suppose that he is quite literally 'no longer himself' while under the supernatural influence of the Siren's song; and maybe also for the psychedelic drugs. But the case seems less straightforward in the more mundane example of the drunken-driver, and there is a danger that the range of counterexamples that van Fraassen would have us dismiss as involving excessive 'disability' will begin to blur into the increasingly contentious. Certainly none of us are at our best after a night on the town, but as Maher (1992: 122) puts it, no one 'gives any real credence to the claim that having 10 drinks, and as a result thinking they can drive safely, would destroy their personal identity'.

This however is where the second horn of the dilemma comes into play. Take a situation where an epistemic agent holds a particular opinion at one time while recognising that he will hold a different and diachronically incoherent opinion at a later time, but where there is no straightforward cognitive impairment that might problematise his personal identity as in the drink, drugs and Homeric temptress examples above. The example van Fraassen (1995a: 24–25) gives is of a scientist who, while firmly rejecting materialism, discovers that belief in materialism is entirely the result of a long-term dietary deficiency that he himself suffers from: thus he currently affords materialism a very low subjective probability; knows that in the future he will afford materialism a very high subjective probability (and for reasons that make it difficult to dismiss such future opinions as no longer being *his* future opinions); and, moreover, seems *prima facie* rational in so doing. Another example, and one that abandons the theme of epistemic agents changing their minds following some kind of ingestion altogether, is provided by Plantinga (1993: 149): I currently have no particular opinion about nominalism, considering it just as likely as its denial; however, I am about to attend the lectures of a well-known nominalist spokesman with a reputation for being a highly persuasive advocate of his views, and therefore know that in the future I will consider nominalism at least twice as likely as its denial. I clearly violate the Reflection Principle, but does it seem *irrational* for me not to consider nominalism twice as likely as its denial right now, *before* I hear the lecturer's arguments?

What the integrity defence draws to our attention however is the fact that there is clearly something amiss with the examples given above: although it inevitably happens, forming an opinion merely on the basis of one's diet, or on the grounds of persuasive rhetoric, is just not part of what our idea of being a responsible epistemic agent amounts to. Were we ourselves faced with just the sort of situation described above, our first response would not involve the calculation of our own diachronic coherence or a re-assessment of our principles of rationality; our first response would be to take measures to remedy our dietary deficiency, or to familiarise ourselves with the arguments against nominalism, so that we do not come to form beliefs on grounds that we do not consider to be appropriately evidential. To do otherwise – that is, not to take any measures for policing how we will go about updating our opinions – is to violate a notion of epistemic integrity vital to our understanding of what an epistemic agent is. As van Fraassen puts it,

> The question about how you will revise your opinion is in the first instance a question about your integrity as epistemic agent. The first reply must be to express your commitment to follow only epistemic policies which you can endorse.
>
> (van Fraassen, 1995a: 25)

In the case of the anti-materialist scientist with the dietary deficiency, the correct conclusion to draw then is not that he violates the Reflection Principle, but simply that 'he will possibly feel more sympathy for materialism, but not automatically equate his inclinations with his considered opinion' (van Fraassen, 1995a: 25). And what if these inclinations are so strong as to render it impossible for him to distinguish between them and his considered opinion? In that case, it would seem fair to say that our scientist's future opinions are no longer in his hands – that is, it is no longer reasonable to think of these future opinions as *his* future opinions – and we are back on the first horn of the dilemma, and the death and disability defence.

The dilemma proposed by van Fraassen is characteristically cunning, although it is difficult to judge whether or not *any* putative counterexample to the Reflection Principle will impale itself upon one horn or the other. Take for example the case of forgetting (Plantinga, 1993: 156–158; Talbott, 1991: 138–142). At the present moment of time, I have an extremely high subjective probability that I saw my friend Nick in the street this morning. However, I also know that at some point in the not too distant future I will have forgotten this, and will assign a rather low subjective probability to having seen Nick at that precise time and place. By the Reflection Principle, this is incoherent; and it appears that I should allow my anticipated future forgetfulness to influence my current credence about what I did this morning. This is clearly an absurd conclusion, but it is hard to see which horn of van Fraassen's dilemma is to come to the rescue at this point. Certainly no mind-altering substances are involved, and there is no sense in which we might maintain that these future doubts are not *my* future doubts. Moreover, it seems rather contentious to claim that my integrity as an epistemic agent requires me to overcome these natural limitations of my cognitive faculties – we forget trivial pieces of information like this all the time, and good thing too lest we overburden ourselves. Of course, a *perfectly* rational epistemic agent may well be one who never forgets anything; but this is not the key in which van Fraassen's advocation of the Reflection Principle is intended.

Clearly a lot more needs to be said about this crucial notion of epistemic integrity. One option may be to argue that when it comes to matters of rationality, not all beliefs are equal: a counterexample to the Reflection Principle must involve only those beliefs to which we attach some importance – our 'considered opinions' about weather prediction, say – rather than just any old belief that we happen to acquire in the process of walking down the street. The fact then that my forgetting when exactly I met my friend seems to be a case of neither cognitive disability nor epistemic vice is of no consequence to van Fraassen's defence of the Reflection Principle, since this is only ever meant to cover those subjective probabilities we deem important to my integrity as an epistemic agent, and we can hardly expect every minutia of our day-to-day lives to be relevant for that.

There will of course be a problem here about where exactly to draw the line between the important beliefs and the not-so-important beliefs, and a risk that any such attempt will simply end up defining the limits of my epistemic integrity so as to rule out counterexamples to van Fraassen's dilemma by fiat. There is however a more important difficulty with van Fraassen's notion of epistemic integrity, at least as far our present discussion is concerned. What epistemic integrity seems to amount to – so far as it is employed in van Fraassen's defence of the Reflection Principle – is that an important part of being a rational epistemic agent is that when one holds a particular belief, one is thereby committed to a particular policy concerning how one might come to update or change that belief. In other words, epistemic integrity is an explicitly future-orientated concept: part of what it is to hold a particular opinion at one moment of time is a constraint on what other opinions one can hold at a later moment of time. Our easily swayed student currently considers nominalism as no more or less likely than its denial; epistemic integrity therefore entails that he cannot subsequently come to consider nominalism far more likely than its denial on the basis of certain procedures for updating his credences. The worry then is that this begins to sound a lot like the voluntarist perspective that van Fraassen's discussion of the Reflection Principle was supposed to *establish*; indeed, as Plantinga (1993: 151–155) suggests, the notion of epistemic integrity that van Fraassen has in mind here is one that explicitly distinguishes between an autobiographical statement of an opinion and an agent's expression of that opinion, emphasises a commitment to maintaining an opinion and understands the making of an epistemic judgement within the larger framework of making other speech-acts. To put the point another way, we can all agree that there is definitely *something* epistemically unsatisfactory about our credulous student calmly awaiting his sudden change of opinion; but it's not at all clear that we would regard any course of action short of avoiding the lecture (or re-calibrating his current opinions about nominalism) as straightforwardly irrational *unless* we were already committed to an epistemology that construes epistemic judgements as something akin to a speech-act.

The situation is in fact rather similar to the diachronic Dutch-book strategies with which we began: at one moment in time, our student considers nominalism as no more or less likely than its denial, and he holds his opinion – let us assume – on a perfectly rational assessment of his current evidence; at a later moment in time, our student considers nominalism as pretty likely, again based upon a perfectly rational assessment of his current evidence (the lecturer may well be particularly persuasive, but he's still giving arguments for his position). In the meantime, the available evidence has of course changed – the arguments presented in the lecture – and from his earlier point of view, there may well be something irresponsible about his having put himself in a position to only hear one side of the story. But the fact remains that each separate set of beliefs is synchronically coherent and evidentially appropriate, and we have no reason to demand diachronic coherence in addition to these virtues unless we are already committed to the voluntarist's future-orientated conception of an epistemic judgement.

2.2.4   Judgements about commitments

There is one final difficulty with arguing for epistemic voluntarism on the basis of the Reflection Principle which is quite distinct from the worries raised so far. Suppose that my foregoing objections miss their target, that van Fraassen is correct to insist that a rational agent must satisfy some constraint of diachronic coherence in his beliefs, that this constraint is captured by the Reflection Principle and that as a consequence one must understand the making of an epistemic judgement as the endorsement of certain commitments. The problem is that, even if we grant all of the above, we are still far from establishing what van Fraassen requires in the context of his constructive empiricism. The motivation here is that if we are persuaded to abandon traditional epistemology in favour of a voluntarist orientation, then we cannot object to the constructive empiricist's regimentation of his scientific beliefs (provided of course that such regimentation is logically consistent) since this would simply be the endorsement of one epistemic policy over another. The line of argument under consideration attempts to secure that epistemological transition on the grounds that since our subjective probabilities must satisfy a constraint of diachronic coherence, we must completely reconceptualise the notion of making an epistemic judgement as the undertaking of a particular course of cognitive action, and therefore abandon the traditional epistemological picture in favour of the voluntarist perspective as the only way in which we can understand this particular aspect of our epistemic lives. What this means however is that in order to undermine the sorts of arguments that we encountered in Chapter 1 – arguments that attempt to show that the constructive empiricist's distinction between the observable and the unobservable is epistemically unfounded – van Fraassen must show that *all* of our epistemic judgements are the endorsements of commitments. There can be no room allowed for the autobiographical conception of an epistemic judgement; for if such a framework can continue to exist side-by-side the voluntarist perspective, we will have no reason to *reject* the traditional picture, and the same objections can continue to be made. Under such an eventuality, all that van Fraassen's argument would amount to would be the somewhat anti-climatic observation that there is a pragmatic incoherence – or perhaps a kind of Moorean performative contradiction – in a rational agent *announcing* certain subjective probabilities in the first-person; but this would be far short of the stronger conclusion that such subjective probabilities were illegitimate *tout court*, and thus far short of establishing the epistemological transition under discussion. But there is nothing in van Fraassen's argument that to make an epistemic judgement is to undertake a particular commitment that precludes a perfectly rational agent from also forming autobiographical beliefs *about* those commitments.

Consider the analogy of promising again. As Plantinga (1993: 158) points out, the act of promising certainly does not usually function as an autobiographical report of an agent's state of mind. But when an agent makes a promise – undertakes a commitment to a certain course of action, etc. – he can clearly note (perhaps even ruefully) that this is what he is *doing*. Such an agent can come to believe the autobiographical statement that he has made a promise just as easily as any third party could form the related biographical belief. When I promise my editor that my manuscript will be finished on time, and she asks me how likely I am to fulfil this promise, van Fraassen is quite right that the correct response is not to cite statistics but simply to reaffirm my commitment. But conversely, when challenged to deliver on a promise I never actually made, the correct response cannot be my continuation in a contrary course of action – we must return to the (autobiographical) facts: I remember that day clearly, and I promised you no such thing.

And if so with promises, why not with degrees of belief? As Plantinga puts it,

> I make whatever commitment integrity requires; I resolve to change belief only in rational ways and for good reasons; but can't I also and quite sensibly have a personal probability for the proposition that I will do what I commit myself to doing? I promise to be faithful to you; if you ask me how likely it is that I *will* be faithful, I can't properly respond by citing statistics about the frequency with which promises of this sort are indeed followed by faithfulness. But can't I nevertheless *have an opinion* (a subjective probability) on the question how likely it is that I will keep the promise?
>
> (Plantinga, 1993: 158)

Of course, there will be some tensions in so doing: if I have a low subjective probability about my commitment to only change or update my degrees of beliefs in a rational manner (I know from experience that I just can't keep away from rhetorically persuasive speakers), this will undoubtedly have some negative impact on my resolve to keep that commitment. Endorsing commitments, and making judgements about those commitments, may well require some epistemic subtlety; we may not be able to do both at the same time. But just as van Fraassen may well be right to emphasise the importance of integrity in managing our epistemic lives, so too is Plantinga right to emphasise the importance of self-knowledge in our higher-order reflections upon that epistemic life. Certainly, nothing that van Fraassen has said rules it out.

The upshot then is this: even if we grant van Fraassen's case for diachronic coherence, and thereby his case for a commitment-based approach to epistemic judgements, all that the argument would establish is that we must *add* voluntarist commitments to our analysis. What the argument must establish however, if it is to do the work required of it in undermining the objections from epistemic relevance that plague the constructive empiricist's distinction between the observable and the unobservable, is that we must *abandon* the autobiographical statement approach altogether. At best, van Fraassen's considerations would *augment* our understanding of epistemic judgements, rather than *replacing* it; and while this may be an important contribution for epistemology in general, it does little work for the philosophy of science.

2.3 Traditional epistemology: False hopes and bad lots

2.3.1 The justification of induction

The second line of argument in favour of epistemic voluntarism is van Fraassen's scepticism at the prospects of a more substantial, traditional epistemology. The point of the argument is to establish that, since one cannot show that any method of ampliative inference is invariably reliable, there can be no question as to what – in general – an epistemic agent is rationally obliged to do. The argument comes in two parts: a specific critique of the method of inference to the best explanation, also known as abduction (van Fraassen, 1989: 142–170); and a more general argument against ampliative inferences *per se* (van Fraassen, 2000). The first argument maintains that since an inference to the best explanation will only be as good as the best explanation under consideration, and since we cannot in general guarantee that our potential explanations will be any good at all, any such inference may well be nothing more than 'the best of a bad lot' (van Fraassen, 1989: 143) and hardly worthy of our rational commitment. The second argument develops the familiar Humean dilemma over the justification of induction, broadly construed: that any such justification will either be deductive or inductive; but since it is not a logical truth that inductive inference constitutes a reliable method of reasoning, and since any inductive justification of induction will simply beg the question, the task of establishing the rational obligation of our ampliative practices is quite hopeless.

We'll begin with the second argument first, since if van Fraassen can establish the irrationality of taking *any* method of ampliative inference as a rule of rationality, his specific criticisms of inference to the best explanation will be rendered somewhat otiose. But before we do so, it is important to specify the following minimal criterion of success – that his arguments amount to more than the mere possibility of radical scepticism. Invoke sufficient evil daemons and envatted brains, and it is possible to undermine any kind of epistemic practice whatsoever. Yet while the possibility of radical scepticism is a serious concern for certain philosophers in the right sort of contexts, it should not be our concern here. What van Fraassen is attempting to establish is that it is straightforwardly irrational to adopt a particular method of ampliative inference as a constraint upon one's reasoning – and that therefore the best policy is to endorse a more relaxed concept of rationality whereby one is *permitted* to make ampliative inferences, and where there is no higher court of appeal beyond one's desire for truth and aversion to error to adjudicate dissenting inferential practices. But epistemic voluntarism is not epistemic anarchy: it is supposed to be a more attractive procedure for balancing our epistemic books, and is itself governed by the basic principle that one should avoid sabotaging one's own prospects for cognitive calibration. It may well be a rather minimal epistemological position, but it is still more substantive than anything vindicated by radical scepticism; we can put the point very simply by noting that if we were under the influence of Descartes' infamous daemon, then even the basic constraints of logical consistency and probabilistic coherence will come to nought, since it is presumably within the daemon's power to condition the world in such a way that even these desiderata can lead us astray.[4] So if we are to take the deceitful daemon hypothesis seriously, then while we may have no good reasons to adopt a method of ampliative inference as a rule of rationality, we also have no good reasons not to do so – for if our entire epistemic existence is to be brought into doubt, one can hardly recommend a permissive conception of rationality as an *improvement* over the traditional epistemological picture, since no epistemic policy will be any better or worse than any other. Or to put the same point another way, radical scepticism may well motivate us to adopt van Fraassen's epistemic voluntarism, but only insofar as if there's nothing to recommend one position over another, we may well prefer his simpler and more austere picture than one that deals with the additional clutter of full-blown epistemological rules. But this is not what van Fraassen's argument is supposed to establish. Epistemic voluntarism is not a default option in the absence of any better alternatives; it is a distinct epistemological framework supported by proper philosophical argument. What all this is to say then is that van Fraassen must proceed cautiously in his case against traditional epistemology – he must present sufficient scepticism to undermine a rule-based epistemology, but not so much scepticism that he renders any discussion of the most appropriate epistemological framework redundant. And as we shall see, this is a very difficult balancing act indeed.

To begin then with the general argument against ampliative rules: as I stated it above, the familiar Humean dilemma over the justification of induction (and in what follows, *mutatis mutandis* for our other ampliative inferences) is that any such attempt will be either deductive or inductive, and neither option holds any hope for success. More broadly, we can see the dilemma in terms of providing either an *a priori* or an *a posteriori* case for the reliability of induction; and this distinction maps onto two successive historical periods of attempted justification, which van Fraassen calls the 'First Way' and the 'Middle Way', respectively (2000: 255–256; the final 'Third Way' is of course reserved for van Fraassen's own voluntaristic dissolution of the problem). Since it cannot be a logical truth that our inductive inferences are reliable, the so-called First Way consisted primarily of attempts to prove *a priori* that our inductive methods satisfied increasingly weaker criterion of adequacy – the prime examples being various pragmatic conjectures, such as Peirce's faith in the eventual convergence of our inductive practices; or the idea often associated with Hans Reichenbach (1938, 1949), that *either* our inductive practices will converge *or* the world is such that it will not admit of *any* method of prediction whatsoever (in which case of course we might just as well continue to make ampliative inferences, since there will hardly be a *better* alternative). Yet even these minimal standards of success are seen to fail. That our inductive practices will converge is of course an empirical question; indeed, such a principle seems to function as much as a guiding heuristic in Peirce's philosophy than it does as a matter of fact.[5] And even our fall-back conditional conjecture that if anything works, induction will work (which amounts to the claim that we can show *a priori* that our methods of inductive inference are reliable *with respect to those domains that we know are amenable to ampliative inference*) seems a step too far. The problem here – and this will be a problem for any *a priori* justification of our ampliative methods – is in essence a problem about reference classes. If our inductive predictions really are going to tend towards a limit for all those cases that will in fact admit of perfect calibration, then they must tend towards a limit for any arbitrary *subsequence* of all those cases that will in fact admit of perfect calibration – we can think of this as the constraint that, for any putatively reliable method of ampliative inference, we should not be able to bring its reliability into doubt merely by asking whether it still applies for *that* domain (van Fraassen, 2000: 259). For example, we would not be terribly impressed with a method for predicting whether or not it will rain tomorrow if we found out that it only works on Mondays. Thus a perfectly calibrated method of inference for a particular domain will be a perfectly calibrated method of inference for any arbitrary sub-sequence of that domain. We can make this constraint a little bit more concrete by stipulating that the arbitrary subsequences in which we are interested must be humanly formulable – this can be cashed out in terms of an effective method – and that the inference rule that is to be cali-brated for these subsequences be humanly usable – again, the formal notion of computability can be used to make this precise. What this conditional criterion amounts to then is the following: if our induc-tive predictions will tend towards a limit for all those cases that will in fact admit of perfect calibration, then for all those domains that do in fact admit of perfect calibration there will a method of ampliative infer-ence that is perfectly calibrated for every computable subsequence of that domain. The problem then is that, with a little mathematical logic, this condition can be shown to be untenable.

The following argument risks becoming rather technical, but the basic result can be demonstrated with the minimum of fuss. Suppose, for the purposes of clarity, that we are interested in whether or not we have a perfectly calibrated method of inference for whether or not it is going to rain: we can take the domain in which we are interested to consist in an infinite sequence of days; and we can specify any arbitrary subsequence of days by simply indicating, for every particular day, whether or not it is a member of that subsequence.[6] Let us write this as a string of '1's and '0's, where a '1' in the nth position of our subsequence indicates that the nth day of our total sequence of days is a member of that subsequence: the subsequence that consists of every other day, for example, would then be written as an alternating sequence of '1's and '0's. Suppose then that our perfectly calibrated method of inference is perfectly calibrated for every computable subsequence – it would then be possible (again, via some effective method) to produce a list of every subsequence for which our method was perfectly calibrated that would *ipso facto* be a list of every computable subsequence. We can now proceed by *reductio* (van Fraassen, 2000: 259–261; this argument is based upon Putnam, 1963). Using the terminology introduced above, we can formulate our list of subsequences for which our method is perfectly calibrated as a series of binary sequences as follows:

	1st Day	2nd Day	3rd Day	4th Day	5th Day	6th Day	→
1	1	1	0	1	0	0	...
2	1	0	0	1	1	1	...
3	0	0	1	1	1	0	...
4	1	1	1	1	0	1	...
↓	...	...	...	...	...	...	...

We now construct the subsequence that differs from the nth subsequence in its nth place for all values of n – this can be done by taking the 1st member of the 1st subsequence, the 2nd member of the 2nd subsequence, the 3rd member of the 3rd subsequence and so on and so forth, and then changing every '1' for a '0' and *vice versa*. From our table above, therefore, our 'diagonal' subsequence would begin '0100'. This subsequence is clearly a computable subsequence, since we have just specified an effective method for constructing it. And clearly, this subsequence will not appear on our list of subsequences for which our method of inference is perfectly calibrated: by construction, the diagonal subsequence differs from *every* subsequence in the list in at least one place. Thus our list of subsequences for which our method of inference is perfectly calibrated is not *ipso facto* a list of every computable subsequence; hence our perfectly calibrated method of inference is not perfectly calibrated for any arbitrary subsequence of the domain in which we are interested; hence our perfectly calibrated method of inference is not actually perfectly calibrated, even for those domains that in fact admit of perfect calibration. Our conditional supposition leads to absurdity, and we are unable to show *a priori* even that our methods of ampliative inference satisfy the modest criterion that if anything works, induction will work.

2.3.2 Reliabilism and scepticism

The *a priori* approach to the justification of induction is thus dealt with somewhat summarily in van Fraassen's discussion, and we can pass over it fairly quickly too since it is with respect to the so-called Middle Way of *a posteriori* justification that most of our present interest lies – not only because the predominant realist strategy for defending our ampliative inferences is an explicitly *a posteriori* one (to be precise, an explicitly *inductive* one), but also because it is here that van Fraassen's delicate balancing act between rules-of-rationality scepticism and full-blown radical scepticism shows signs of wobbling. The essence of van Fraassen's objection is that one cannot appeal to one's empirical knowledge in order to justify the ampliative inferences upon which that knowledge is based, not because such a strategy is circular (it is after all an explicitly *a posteriori* approach), but rather because our empirical knowledge is actually rather self-conscious about its own limitations and fallibility. In short, it is not that an inductive justification of induction is methodologically unsound, but simply that any such induction will lack the evidential basis for a convincing inference. Given however that such strategies are primarily concerned with the reliability of induction, rather than our knowledge of the reliability of induction, it remains a delicate issue the extent to which such a defence need take seriously van Fraassen's pessimistic pronouncement.

Inductive justifications of induction centre around two crucial premises: that an inferential practice need not be *necessarily* reliable in order for us to be justified in using it; and that we need not *know* that an inferential practice is reliable in order for us to be justified in using it.[7] An inferential practice is necessarily reliable if it tends to draw true conclusions from true premises in every possible world; it is reliable *simpliciter* provided it tends to draw true conclusions from true premises in the actual world. The opening move then for an inductive justification for induction is to lower the standards for what constitutes a justified inferential practice – actual success, rather than logical infallibility. Not only is this an independently plausible assumption to make – we are after all interested in the inferences that we actually perform – this re-conceptualisation of the criteria for success also has important consequences for the evidence relevant for its assessment. An important part of the Humean dilemma is the claim that since we can *conceive* of our inductive inferences going badly wrong – in other words, that there is a possible world where our inferences draw false conclusions from true premises – such practices stand in need of justification. But this is to presuppose that in order to justify induction one must show it to be infallible. Yet the mere logical possibility of induction leading us astray is simply irrelevant to the more modest question as to whether or not it is *actually* reliable. Consequently, one need not take the mere possibility of error as bringing the justification of induction into doubt – only the actual track record of our inductive practices will be relevant to our evaluation.

The move from infallibility to reliability therefore changes the evidential basis for the justification of induction. The second key premise in the strategy under consideration – that in order for an inference to be reliable, one need not know that inference to be reliable – is what allows the inductive inference to the reliability of induction. The basic idea is to distinguish between explicitly *appealing* to the reliability of our inductive inferences as a premise in our argument (which would clearly beg the question) and merely *using* the reliability of our inductive inferences as the way in which our argument proceeds. All that has to be the case is that induction *is* reliable; and given this supposition (which we need not be in a position to know), to infer the future reliability of our inductive inferences on the basis of the past reliability of our inductive inferences will be entirely above board. In the jargon, by adopting this externalist epistemology, the strategy under consideration need only be considered *rule-circular* (merely using the inferential method under discussion) rather than *premise-circular* (explicitly appealing to the inferential method under discussion).

Putting these two moves together then, we first note that the actual track record of induction is pretty good (unlike, to be sure, its possible track record). We then argue that since we do not require it to be *known* that induction is reliable in order for it to be reliable, and since we no longer count the mere possibility of error as evidence against it, we are in as good a position as we could possibly be to infer the future reliability of our inductive inferences on the basis of the past reliability of our inductive inferences. We thus have an argument for the reliability of induction which – while itself an instance of an inductive inference – neither explicitly presupposes that which is to be shown (it is not a premise of the argument that induction is reliable) nor fails to satisfy our (explicitly externalist) standards of justification.

The general strategy can also be illuminated in terms of the distinction between *answering* the inductive sceptic and *pre-empting* him. If we took the criterion for inductive reliability to be logical infallibility, then the mere possibility of error would bring our inductive practices into disrepute, and we would indeed have to provide some justification of induction that did not employ (let alone appeal to) the reliability of induction. However, on the present proposal, the criterion for inductive reliability is only that it tends to work in the actual world: thus the mere possibility of error is irrelevant for our concerns, and our inductive practices – rather than being brought into disrepute – remain available for whatever justificatory needs we have of them. By rejecting Hume's implausibly strict standards of success, and denying that the reliability of induction needs to be epistemically accessible, one can deprive the sceptic of his starting point and approach the inductive justification of induction as a benign instance of a self-supporting methodology.

The important lacuna in this strategy, and the element with which van Fraassen takes issue, concerns the actual track record of induction.

Grant all that has been said above about the unrealistic presuppositions that drive inductive scepticism, the harmlessness of rule-circularity and the virtues of an inductive justification of induction in general; securing such a methodology will be for nought unless the evidential basis really does show that induction has been a reliable guide in the past. And this van Fraassen (2000: 264–268) denies – our empirical evidence attests to the fact that our ampliative inferences are only successful under very specific circumstances, and our ampliative inferences cannot in principle tell us what those circumstances are. As van Fraassen points out, our most ubiquitous examples of an ampliative inference are racial stereotyping and short-range weather prediction, the latter barely more reliable than the former (2000: 264–265); and if our empirical knowledge tells us anything about the conditions under which an ampliative inference will be reliable, it tells us that such knowledge cannot be provided by the inference in question:

> If we use induction ... it sometimes works and sometimes does not. Can induction tell us when this sort of extrapolation will succeed and when it won't? This is the place where science has something to tell us: if science is true, success will depend on facts of microstructure and cosmic structure which cannot be among the inputs for human induction. So the answer is No: induction cannot tell us which applications of induction will succeed.
>
> (van Fraassen, 2000: 266)

Goodman's (1954) so-called new riddle of induction challenges the evidential basis for an inductive justification of induction by showing that in the absence of a principled distinction between our projectable and non-projectable predicates, our inductive track record can be taken as providing evidence for just as many failed inferences as it does for successful; van Fraassen's (2000: 266–271) point is the more concrete one that since our empirical knowledge is conditional upon various boundary conditions, background assumptions and approximation principles – all of which our empirical knowledge tells us lie outside the scope of the ampliative inferences that such knowledge was meant to support – we simply do not have a sufficient evidential basis for inferring the reliability of induction, *a posteriori* or otherwise.

As I noted at the outset, the committed sceptic will find much to his dissatisfaction in any attempted justification of our inferential practices. This is particularly telling in the case of an inductive justification for induction, which is essentially premised on the notion that one should consider induction innocent until proven guilty (contrast the two approaches to scepticism discussed in Chapter 1). Thus, for those already firmly in the grip of their Cartesian doubts, these kinds of *a posteriori* manoeuvres will never even get themselves off the ground. But as I also noted at the outset, this is not the sort of objection that van Fraassen is meant to be offering – any criticism of the inductive justification of induction that depended upon such radically sceptical considerations would similarly undermine his own alternative epistemic framework.

So how should we assess van Fraassen's worries here? As the exposition above makes clear, van Fraassen is happy with the idea that one need not know that induction is reliable in order for it actually to be reliable; and he is happy with the idea that, in the context of a less demanding criterion for justification, this means that our inductive practices may well be available for us to use in such justificatory endeavours. His objection is that we have empirical grounds for believing our inductive inferences to be reliable only under specific circumstances, and that since we are unable to discover these circumstances via our inductive inferences, we can't be sure that our inductive inferences *really are* reliable, epistemological externalism and *a posteriori* methodology notwithstanding. But it was never part of the defence that our inductive inferences are invariably reliable, merely that their actual use tends to produce true conclusions from true premises; and it was certainly never part of the defence that we need to *know* that any particular use of induction is reliable. So while he may be perfectly correct to point out that we do have hard-won empirical evidence for the occasional shortcomings of our ampliative practices, van Fraassen can only develop this into an actual objection to the inductive justification of induction by either demonstrating that our bad inductions significantly outnumber our good inductions or by demanding that any future inductive inference only be performed in the knowledge that this is a circumstance in which it will work. Yet in the first case (at least, barring substantial further investigation) we are in conflict with the explicitly conceded weaker criterion of actual reliability that underlies the *a posteriori* methodology; and in the second case we are in conflict with the explicitly conceded epistemological externalism that supports the crucial distinction between vicious premise-circularity and benign rule-circularity. Consequently, it's hard to see exactly what van Fraassen's observation amounts to, beyond the basic sceptical disposition that we dismissed above.

The idea of course is that since it is *empirical* evidence that van Fraassen is pointing to – what our scientific knowledge tells us about the limitations of those ampliative inferences upon which they are based – these are concerns that any *a posteriori* justification of induction must take far more seriously than the mere logical possibilities that drove Hume's dilemma. And this is certainly true enough, as far as it goes. The problem remains however that, despite its greater pedigree, van Fraassen's evidence still only amounts to the observation that induction is not infallible and that we don't know very much about when and where it will be successful. But since these are two requirements that the inductive justification of induction explicitly denies that it needs to meet, the strength of van Fraassen's objection lies merely in the degree of scepticism we wish to bring to the project – scepticism that threatens to undermine his own, alternative, epistemological framework.

2.3.3   The 'best of a bad lot'

The general argument against the justification of our methods of ampliative inference is therefore unsuccessful. More specifically, while we may indeed remain sceptical about any putatively inductive justification of induction – we may query the reliabilist criterion of adequacy, or reject the externalist epistemology that such a justification depends upon – the problem is that we cannot also rely upon such scepticism to provide a positive motivation for epistemic voluntarism. In short, the way in which we may feel that the *a posteriori* justification of induction falls short of its goal is that it says nothing in response to the committed sceptic – rather, to return to an earlier distinction, it allows those with an inductively favourable disposition to successfully *pre-empt* the sceptical challenge; it shows how those of us already pre-disposed to reason inductively can construe their position in such a way so as not to be unduly concerned by the familiar Humean dilemma. But to demand more than this from our justificatory practices is to try and meet the sceptic head-on; and while this may indeed be a laudable challenge, such a radically sceptical context will not allow us to motivate epistemic voluntarism over traditional epistemology, since such radical scepticism will equally undermine the (admittedly, more parsimonious) tenets of a more permissive rationality.

Let us turn then to van Fraassen's more specific criticisms of inference to the best explanation. Although a narrower target, inference to the best explanation is in many ways the more important inferential practice for van Fraassen, since it is the primary method of reasoning deployed in arguments in favour of scientific realism – that the *best explanation* for the success of our scientific theories is that they are (at least approximately) *true*. Even in the absence then of a general case against ampliative inference – and therefore in the absence of a general argument for epistemic voluntarism – a successful attack on inference to the best explanation would at least motivate the plausibility of a constructive empiricist approach to science, as outlined in Chapter 1. Unfortunately, as we shall see, the same instability between applying sufficient scepticism to undermine a rules-based rationality and applying so much scepticism as to undermine the prospects for any alternative epistemic framework similarly de-rails the case to be made here.

In order to grasp van Fraassen's objection, it is useful to think of an inference to the best explanation as involving a two-stage process: during the first stage, various potential explanations for the phenomena in question are considered and ranked according to our various criteria of loveliness, likeliness and so forth;[8] and in the second stage, whichever potential explanation has made it to the top of our list is then inferred to be true (i.e., considered to be an *actual* explanation). The basic difficulty that van Fraassen (1989: 142–149) finds with this process concerns the ranking of hypotheses during the first stage. For even if we suppose that we have well-established criteria for what makes one potential explanation better than another, and even if we suppose that we are perfectly reliable in applying these criteria, the fact remains that we can only rank those potential explanations that we have actually considered. Yet for any phenomena that we wish to explain, there will be an infinite number of empirically adequate (yet mutually incompatible) potential explanations, only a vanishingly small number of which will have been compared and contrasted. The worry then, in van Fraassen's memorable phrase, is that 'our selection may well be the best of a bad lot' (1989: 142). In order to have any confidence that our best potential explanation is indeed the actual explanation (that it is indeed true), we need to have some justification for thinking that the truth will be among the tiny subset of potential explanations that we have actually considered. Yet this is to assume an awful lot about our intellectual practices – why should our attempts to propose potential explanations enjoy such an epistemological privilege? Although initially tempting, inference to the best explanation therefore seems a hopelessly optimistic policy. If van Fraassen is right then, as Lipton puts it, 'on this view, to believe that the best available theory is true would be rather like believing that Jones will win the Olympics when all one knows is that he is the fastest miler in Britain' (Lipton, 1993: 90).

This then is the essence of van Fraassen's so-called 'best of a bad lot' objection to inference to the best explanation, and the rest of his discussion is an attempt to pre-empt any putative response. In order to salvage inference to the best explanation, one would have to provide some kind of justification for the assumption that the actual explanation will always lie among the range of potential explanations under consideration, and it is very difficult to make a compelling case for this degree of epistemic privilege (van Fraassen, 1989: 143–145).[9] The most promising proposal in van Fraassen's view will be a matter of retrenchment – inference to the best explanation is not to be construed as the policy of inferring the truth of our best explanation, as that presupposes an implausibly optimistic epistemic self-assessment; rather, the view is merely that explanatory virtues have an important role to play in determining which beliefs we come to adopt. But more needs to be said on the role these explanatory features play, and here we face a dilemma. On the one hand, we could argue that explanatory virtues such as simplicity and strength are guides to the truth; and since these features will heavily influence which potential explanations we come to consider, our ranking of hypotheses does operate under some moderate degree of privilege. On the other hand, we could argue that explanatory considerations contribute directly to our distribution of subjective probabilities – that the degree of belief that we should have in a proposition is a function of both our conditionalisation upon the evidence *and* the explanatory weight of that proposition. Yet on the first – purely *qualitative* – account of the importance of explanatory considerations, we may well wonder just why it is that simplicity is a guide to truth (van Fraassen, 1989: 146–148). And on the second – *quantitative* – account, giving a probabilistic boost to explanatory hypotheses beyond straightforward conditionalisation will simply violate the probability calculus and engender flat-out irrationality (van Fraassen, 1989: 160–170).

The Bad Lot objection does offer a genuine challenge for any advocate of abductive reasoning, and in particular, those who would wish to reason abductively for realism in the philosophy of science. But before we do begin to assess van Fraassen's case against inference to the best explanation, it is important to re-iterate what exactly is his target here. A common complaint levelled against van Fraassen's argument is that if it were successful, it would apply equally to his own constructive empiricism and would therefore be self-undermining, failing to constitute a case in favour of one position over the other. As Psillos (1996: 41; the following interpretation is also endorsed in Psillos, 1999: 211–215) puts it, the inference that a successful scientific theory is empirically adequate requires us to go beyond our available evidence just as the inference that a successful scientific theory is true (an empirically adequate theory, recall, is one that gets it right about *all* of the observable phenomena – past, present and future). Consequently, the constructive empiricist can only *infer* that his best scientific theory is empirically adequate, just as the scientific realist can only infer that such a theory is true; and if the scientific realist must justify his assumption that the actual explanation will lie among the considered explanations, so too must the constructive empiricist justify his assumption that the empirically adequate explanation will also lie within this set.

As Psillos sees it then, van Fraassen's objection cannot be to the general practice of inference to the best explanation, since his own philosophy of science depends upon such a method of reasoning. Rather, van Fraassen must object to the particular *species* of inference to the best explanation deployed by the scientific realist – an inference to the existence of unobservable entities (i.e., the truth, rather than the empirical adequacy, of a scientific theory), which Psillos describes as a case of 'vertical' inference to the best explanation. This is to be contrasted with the species of inference to the best explanation allegedly favoured by the constructive empiricist which merely infers the existence of observable entities – 'horizontal' inference to the best explanation, which only takes us as far as the empirical adequacy of our best scientific theory. In essence, while a horizontal inference only leads to the postulation of more of the sorts of observable entities that we already believe in, a vertical inference takes us into a new domain and thereby posits different *types* of entities – in this case, unobservable ones. The problem then, according to Psillos, is that this distinction fails to capture anything of epistemic relevance – both horizontal and vertical inferences must presuppose a degree of privilege to justify the assumption that the best theory we have considered is the best theory *simpliciter*; and if the constructive empiricist is entitled to such an assumption, then so too is the scientific realist.

In many ways then, Psillos' argument can be seen as a generalised version of the sort of complaint that we encountered earlier in Chapter 1. For example, according to Churchland (1985: 39–41), van Fraassen's willingness to count unobserved, spatio-temporally inaccessible, yet macroscopic objects as observable-in-principle, while refusing to count unobserved, spatio-temporally proximate, yet microscopic objects the same, was indicative of epistemic double-standards – what grounds our faith in the reliability of what our scientific theories conjecture about big things, but not for what they conjecture about small things? After all, both require counterfactual leaps of faith, and both have been vulnerable to revision. Such a line of thought finds easy expression within Psillos' distinction between horizontal and vertical species of inference to the best explanation – one can believe what our scientific theories say about an unobserved moon, for although it will involve some kind of abductive step, it is an abductive step that only leads to more observable objects (thus a horizontal inference); by contrast, one cannot believe what our scientific theories say about unobserved electrons, since this would involve abducting to the existence of something beyond the remit of our empirical evidence (a vertical inference).

The connection with Churchland is instructive, for it reveals a similar inadequacy in Psillos' argument. In the former case, we have seen that Churchland rejects the idea that our observable ontologies are more epistemically secure than our unobservable ontologies on the grounds of comparable theoretical infection. The problem with the argument is that to be effective, one must establish the *quantitative* claim that our observable ontologies are just as theoretically infected as our unobservable ontologies; yet all Churchland establishes is the *qualitative* claim that our observable ontologies are theoretically infected, just as our unobservable ontologies are. Similarly, as Ladyman et al. (1997: 307–308) point out, although Psillos may be able to establish that both scientific realists and constructive empiricists require some assumption of background privilege to run their respective inferences, he fails to establish that they will require the same degree of background privilege. For whereas the scientific realist needs to guarantee that the true hypothesis will be among those he considers, the constructive empiricist merely needs to guarantee that an empirically adequate hypothesis will be included; and since there will be an infinite number of empirically equivalent alternatives for any empirically adequate theory, yet only one true theory, it looks as if the degree of privilege invoked by the constructive empiricist will be significantly weaker than that of the scientific realist.

We can however safely leave to one side this particular line of argument, for *contra* Psillos – and by extension, *contra* Churchland, Hacking and all the other critics who claim to discern a pernicious inferential discrimination underlying constructive empiricism (although, again, we have seen reasons to doubt their case) – van Fraassen's critique of inference to the best explanation does not rest upon a distinction between horizontal and vertical aspects of this method of reasoning, and nor does it presuppose a greater epistemological security in the claims we make about observable entities. The critique is meant to concern inference to the best explanation in general, and not to specific applications. Indeed, our presentation of the 'best of a bad lot' objection at the beginning of this section made no such distinction as to its range of application; and even if one can argue against Psillos that an inference to the empirical adequacy of our best theories is an easier task than an inference to their truth, one must still say *something* about why our ranking of hypotheses enjoys even this much privilege. Yet Psillos is right when he notes that van Fraassen remains content to infer that his scientific theories are empirically adequate, and that this may seem perverse without a proper appreciation of what the Bad Lot argument is meant to achieve.

In order to help clarify the situation, let us return to one of van Fraassen's earlier arguments over the application of abductive reasoning in the philosophy of science, one which Psillos (1996: 34; 1999: 213–214) cites in support of his two-inferences interpretation. I hear the patter of tiny feet in the night, and my cheese disappears from the fridge; I infer that a mouse has come to stay (van Fraassen, 1980: 19–21). Yet I need not assume that what I have deployed here is an inference to the *truth* of my best explanation, since my argumentative pattern is consistent with an alternative inference to the *empirical adequacy* of my best explanation – the distinction is often missed, because when we are discussing observable phenomena, inference to the truth and inference to the empirically adequate will coincide. Crucially then, when the scientific realist points to the tracks in a cloud chamber and the readings on the meter, I need not follow his lead and infer the truth of his electron hypothesis – I may instead infer that things are *as if* the electron hypothesis is true, and no unobservable entities need enter my ontology.

A natural interpretation of this argument is that van Fraassen does indeed distinguish between abductive inference to the empirical adequacy of a theory and abductive inference to its truth; that he does maintain that abductive inference to the empirically adequate is more reliable or epistemically secure than abductive inference to the truth; and that in his view scientific realists often misconstrue evidence for the reliability of the former as evidence for the reliability of the latter, since when we are only talking about observable phenomena, an inference to the empirically adequate is equivalent to an inference to the truth. But in contrasting abduction to the truth with an abduction to the empirically adequate, van Fraassen is not intending to introduce an alternative method of reasoning; rather, abduction to the empirically adequate is intended as a *counterexample* to the claim that the only interpretation of our everyday scientific reasoning is as an inference to the truth of our best explanation (Ladyman et al., 1997: 313–314). The basic argument for scientific realism often proceeds by citing the success-ful application of inference to the best explanation in the construction of scientific theories, arguing specifically that these are instances of an inference to the truth of our best explanations, concluding that inference to the truth of our best explanations is a reliable method of reasoning, and then employing such an inference to move to scientific realism in general. The basic point of the mouse example is that we can-not automatically assume that the successful application of inference to the best explanation is an instance of inference to the *truth* of our best explanations, since the same inferential phenomena admit of countless other interpretations, such as inference to the empirical adequacy of our explanations.

Once we focus on this idea of generating counterexamples to what the scientific realist claims, rather than advocating an alternative epis-temological strategy, we can understand how exactly van Fraassen can continue to endorse his own abductive inferences while apparently crit-icising the practice in general. What van Fraassen wants to undermine is not the practice of inferring the truth of our best explanations, but in understanding such a method of reasoning as a *rule* – that is, a prin-ciple that one is always obliged to follow. What the Bad Lot argument is meant to show is that one cannot always guarantee that one's best explanation is *the* best explanation; consequently, one cannot always suppose that such an inference is the most rational step to take. Of course, if one *does* believe that the actual explanation lies among the potential explanations under consideration, then an inference to that best explanation is perfectly acceptable – it is 'a case of common sense which no-one will deny' (van Fraassen, 1989: 149). But only in those cir-cumstances; and what the Bad Lot objection is supposed to make clear is that such favourable circumstances may be few and far between. To put the point another way: in lieu of a convincing general response to the Bad Lot objection, we must recognise that an inference to the best expla-nation is only rational in those contexts in which we can reasonably believe the truth to be among the considered hypotheses; consequently, inference to the best explanation cannot be a *rule* of inference which we are always obliged to follow, since we may well be in an unfavourable context. What remains then is a *choice*: use the principle as widely or as narrowly as your desire for truth and aversion for error demand. This may license belief in the truth of one's scientific theories, or it may only license belief in their empirical adequacy; the only thing that one must avoid is to suppose that one's own policy makes rational demands upon others. As van Fraassen puts it,

> Someone who comes to hold a belief because he found it explanatory, is not *thereby* irrational. He becomes irrational, however, if he adopts it as a rule to do so, and even more if he regards us as rationally compelled by it.
>
> (van Fraassen, 1989: 142)

2.3.4 Background theories and the ranking of hypotheses

The essence of van Fraassen's Bad Lot objection is therefore the claim that, since one cannot guarantee that the actual explanation will always lie among the potential explanations under consideration, one cannot be rationally *obliged* to abduct upon one's evidence. Inference to the best explanation cannot therefore be a *rule* of rationality; and if one holds similar opinions about our other methods of ampliative inference, traditional epistemology is rendered empty. At the least, those aspects of our inferential practices that are supposed to constitute a case for scientific realism are rendered permissible rather than obligatory. Consequently, there are no criteria beyond the simple expediency of avoiding epistemic self-sabotage to adjudicate our choice of epistemological posture; and no criteria beyond logical consistency and probabilistic coherence in our decisions to believe a hypothesis to be true because it is explanatory, or to limit our interests to the empirically adequate.

The obvious response to the Bad Lot objection then is to argue that we can in fact guarantee that the best explanation under consideration is the best explanation *simpliciter*. There are two ways we might do this. The first is by exhaustion – to argue that when we rank the various potential explanations under consideration, we are in fact ranking all the possible potential explanations. As Lipton (1993: 93–95) points out, this need not necessitate the utterly unfeasible task of considering an infinite number of distinct hypotheses. Rather, all that one needs to do is to make sure that one's considered hypotheses exhaust the logical space of competitors by including the negation of one's considered hypotheses. The simplest case then would be to consider the hypothesis H and its negation ¬H, which as a matter of logic will exhaust all potential explanations of the phenomena (all the other hypotheses can be characterised *en masse* in terms of the denial of H).

The problem of course with such a strategy is that one may quite rightly object that ¬H is not much of an explanation: suppose we had

several potential explanations available for some particular phenomena, and that after careful consideration we decided that the negative
hypothesis – the best explanation is: none of the others! – satisfied
better whatever criteria we had in play. For the argument by exhaustion to work, we would have to be happy with an explanation that
simply stated that all of the alternatives we had thought of were not
likely to be true, rather than chalking up such an eventuality as the *failure* to come up with a compelling explanation for the phenomena in
question.

In any case, there is a second strategy for reassuring the would-
be abductor of the reliability of our explanatory considerations which
points to a tension at the very heart of the problem. As I outlined it
above, the Bad Lot objection is based on the following two premises:
that we can reliably rank the various potential explanations under
consideration, but that we have no reason to suppose that the set of
potential explanations under consideration is likely to include the truth.
Following Lipton (1993: 95–101) let us refer to these as the Ranking
Premise and the No-Privilege Premise, respectively. It is important to
note why exactly van Fraassen must grant the Ranking Premise in his
discussion, that is, to concede for the sake of argument that we can
reliably rank the potential explanations under consideration. This is
not simply the argumentative strategy of attacking one's opponent at
his strongest position. Rather, van Fraassen must concede this much
to his opponent in order for his argument to amount to more than
the mere possibility of radical scepticism. It would be easy to cast
doubt upon our abductive inferences by challenging whether or not
we can reliably rank the potential explanations under consideration,
by querying the correlation between those potential explanations we
favour and those potential explanations that are true or by simply
contesting our capacity to successfully understand the external world
at all. Yet just as with van Fraassen's critique of our ampliative inferences in general, while this may indeed be a pertinent issue for some
philosophers in certain contexts, it cannot be what the Bad Lot objection amounts to. To deny the Ranking Premise would be to raise a
spectre of scepticism that would undermine any potential abductive
inference – inference to the best explanation would not only fail to
an obligatory inferential practice, it would also fail to be a permissible
inferential practice. Such an argument would indeed undermine van
Fraassen's constructive empiricism just as readily as his opponent's scientific realism, just as Psillos alleges. Or to make the parallel with van
Fraassen's general critique of ampliative inference explicit, an argument

that began with such a degree of scepticism may well undermine a rules-based rationality; but since it would also undermine any alternative epistemological framework – no matter how parsimonious – it could hardly motivate van Fraassen's epistemic voluntarism as an improvement.

The delicate balancing act between applying sufficient scepticism to unbalance his opponent and not applying so much as to unbalance himself manifests itself in the Bad Lot argument by conceding a certain degree of epistemological success to our would be abductor – whatever the content or origin of the potential explanations under consideration, we can at least trust ourselves in their relative evaluation. But just as granting the inductivist his reliabilist criteria of success (and his externalist epistemology) *ipso facto* rendered van Fraassen's more specific misgivings about the limitations of induction redundant, so too does granting the abductivist his reliable ranking of potential explanations render van Fraassen's general challenge about the privilege of this set of explanations toothless. As Lipton (1993: 95–101) notes, the ranking of potential explanations does not take place in a vacuum: the various background theories that we accept will play an important role in our relative assessments.[10] But if we grant that our ranking of potential explanations is a reliable process, and we concede that such ranking depends heavily upon our accepted background theories, we must also grant that our background theories must enjoy some privilege or success in order to have played such a role. So if we grant the Ranking Premise, we must also grant the approximate truth of our background theories. But these background theories were themselves the product of earlier inferences to the best explanation; it is therefore the case that our previous reliable rankings of potential explanations did indeed tend to select the best explanation *simpliciter*. We can therefore conclude (with a little induction which, fortunately, we have not seen any reason to bring into doubt) that our set of potential explanations does tend to enjoy the necessary degree of privilege and include the actual explanation. Grant the Ranking Premise, and it is no longer coherent to maintain the No-Privilege Premise too. Thus just as with his general case against ampliative inference, van Fraassen's case against abductive reasoning fails to strike the appropriate sceptical balance.

2.3.5 Probabilistic and non-probabilistic inferences

In van Fraassen's presentation of the argument, the issue as to whether or not drawing inferences on the basis of explanatory considerations leads to probabilistic coherence is very much a secondary concern. The principle objection is that we have no good reasons to suppose that the actual explanation will lie among the potential explanations that we consider, and that thereby an inference to the 'best' explanation may not be very good at all. The argument concerning probabilistic coherence is directed at one possible retrenchment the embattled abductor may attempt – to switch from inference to the best explanation as normally conceived, and defend instead an essentially Bayesian approach which nevertheless incorporates an important role for explanatory considerations. However, since I have argued that van Fraassen's initial objection fails (relying as it does upon a particularly corrosive form of scepticism), it may seem that we can thereby ignore this part of the dialectic altogether: if one need not retrench in one's response to the Bad Lot objection, one need not switch to a potentially problematic variation of Bayesianism. Nevertheless, one can easily motivate a closely related and quite independent objection to abductive reasoning based upon the probabilistic considerations that van Fraassen raises. The objection would be not so much that any putatively abductive inference must take the form of an explanatorily motivated Bayesian conditionalisation; rather, since Bayesian conditionalisation provides the basic constraints of probabilistic coherence, any ampliative inference must be shown to be consistent with this mechanism regardless. But on first inspection, since an inference to the best explanation will take into account various features that go beyond the basic acquisition of new evidence – such as how much explanatory weight that evidence may provide – it looks as if such a method of inference will be probabilistically incoherent, whether or not it requires retrenchment.

Let us develop the worry a little more slowly. According to the basic Bayesian formula, our subjective probability for a hypothesis H, given the acquisition of a piece of evidence E (our 'posterior probability' for H given that E), should be a function of our prior subjective probability for H (independently of this new piece of evidence E), our prior subjective probability for E (independently of our hypothesis H) and our prior subjective probability for how likely E would be to occur if H were the case. More formally:

$$P(H/E) = \frac{P(E/H).P(H)}{P(E)}$$

Very intuitively, if our prior subjective probability for H is rather low (it is a new conjecture in which we have little faith), and our prior subjective probability for E is rather low (it is an unlikely phenomena that we have no independent reason to expect) but our subjective probability for E occurring if H were true rather high (because, for example, H predicts E), then upon the acquisition of evidence E, our posterior subjective probability for H will increase significantly. And this is how it should be: in the case described above, E is excellent evidence for H; and since we have no other good reasons to expect E to occur, all this should increase our confidence that H is true.

Bayesian conditionalisation therefore captures many of our basic intuitions concerning confirmation. Moreover, it is easy to show that any distribution of subjective probabilities that did not conform to this formula would be vulnerable to the usual detrimental betting strategies: if one's posterior subjective probability for H given E was not a function of one's prior subjective probability for H, one's prior subjective probability for E and in one's prior subjective probability for E given H, then the cunning bookie would be able to offer you one wager based on your posterior subjective probability, and another wager based on your prior subjective probabilities which, although they would both concern the same state of affairs, would be of different values. And note that nothing about this strategy would involve the contentious issue of diachronic coherence: to hold a posterior subjective probability for H given E that was not a function of one's prior subjective probability for H would be as straightforwardly *synchronically* incoherent as believing that A, believing that if A then B, but not believing that B.

The problem then for inference to the best explanation is as follows. Since the whole point of an inference to the best explanation is to link our ampliative reasoning to some concrete theoretical virtue – that is, to how good an explanation that hypothesis would be – and to thereby give our inferences more focus than straightforward enumerative induction, our erstwhile abductor seems therefore committed to the policy that a good explanation *counts for more*: faced with two competing hypotheses both equally supported by the data, we should still consider the one that provides the better explanation to be more likely to be true. Yet once we allow explanatory virtue to provide this kind of boost to our subjective probabilities, we have clearly violated Bayes' Theorem. To return to our example using *modus ponens*, it would be like believing that A, coming to believe that if A then B, but on the grounds that it carries little explanatory weight, refusing to believe that B. It seems then that since its very essence is to provide more content than a straightforward enumerative induction, inference to the best explanation is probabilistically incoherent *simpliciter*.

There are however a number of ways to tackle this problem. The first is to note that while Bayesian conditionalisation does impose important constraints on how one updates one's distributions of subjective probability, these constraints may not be as strong as they first appear. As Lipton (2004: 106; cf. Howson, 2000: Ch. 7) notes, while Bayesian conditionalisation imposes a certain *structure* upon our beliefs, it says nothing about the *content* of those beliefs; or in other words, while it may tell us which combinations of beliefs/distributions of credences are incoherent, it doesn't tell us which beliefs/credences must be sacrificed. There is nothing logically inconsistent in believing that A, coming to believe that if A then B, and refusing to believe that B *provided* one is thereby willing to give up one's belief that A (coming to believe the conditional in effect serves as a *reductio* of one's previous beliefs). Similarly, there is nothing probabilistically incoherent in adopting an extra-high subjective probability for H given E on explanatory grounds, provided one is also willing to retrospectively recalibrate one's prior subjective probabilities. So while the constraints of Bayesian conditionalisation show us that one cannot boost one's credences in one's posterior subjective probabilities without thereby applying these explanatory considerations across the board, they do not show that explanation-based inference is probabilistically incoherent in principle.

The idea then is that theoretical virtues like explanatory value can guide our ampliative inferences without violating the probability calculus, provided that such considerations are coherently distributed across our web of beliefs. Lipton (2004: 107–117) generalises this thought: he argues that not only might probabilistic considerations constrain our explanatory practices, but that our explanatory practices might help us to implement our probabilistic reasoning. It goes without saying that while we agree that our distributions of subjective probabilities should satisfy the probability calculus in the way formulised by Bayesian conditionalisation, we are in general very bad at realising this desideratum (Lipton, 2004: 108–109; Lipton here draws heavily upon Kahneman et al., 1982). In general, we need to follow various non-probabilistic heuristics – such as going with the best explanation, etc. – if we are to have any hope of avoiding complete doxastic incoherence. His claim then is that rather than being in conflict, probabilistic and non-probabilistic methods of inference can in fact be seen as two sides of the same coin.

Specifically, we can see how an explanation-based approach may be ineliminable to Bayesian conditionalisation in the following ways.

Firstly, while the probability calculus tells us how to determine our posterior subjective probability for H given E, it tells us nothing about how to determine our prior subjective probabilities for H or for E, and it tells us nothing about how to determine our prior subjective probabilities for how likely E would be were H to occur. In other words, while Bayesian conditionalisation may tell us how to coherently structure our existing credences, it is completely silent on what those existing credences might be. In those cases then where the hypothesis H does not logically entail the evidence E (where of course our subjective probability for E given H should equal 1), we may well appeal to how well H would explain E in determining our subjective probability (Lipton, 2004: 114–115). And similarly, while of course many of our prior subjective probabilities for Bayesian conditionalisation will be the posterior subjective probabilities of a previous conditionalisation, a natural place for this process to start would be by assigning our starting credences on the basis of how well the hypotheses in question explain our current background beliefs (Lipton, 2004: 115–116). Secondly, not only does the Bayesian formula tell us nothing about the input values for our conditionalisations, it also tells us nothing about *when* such conditionalisation should occur. When a new piece of evidence comes to light, we should update our distribution of credences – but what counts as a piece of evidence? Even leaving aside such thorny issues as the proper individuation criteria for events, we may worry that since any aspect of the world will have *some* relevance for any hypothesis, the Bayesian account will leave us constantly tweaking our subjective probabilities in the light of anything that transpires.[11] Yet this is clearly not what happens (and is clearly not what should happen, either); one natural rejoinder then is to note that our selection of what counts as a relevant piece of evidence, and our decision upon when our credences need updating, will appeal to some non-probabilistic considerations – such as when a piece of evidence has particular explanatory consequences (Lipton, 2004: 116–117).

As Lipton happily notes, some of these ideas are somewhat sketchy. The point however is that inference to the best explanation need not be understood as a probabilistic method of inference that affords extra value to those hypotheses that meet certain theoretical virtues, and which thereby lands its practitioner in probabilistic incoherence. One can favour those hypotheses that provide better explanations provided the consequence of these non-probabilistic factors is properly distributed among one's web of beliefs as a whole; and moreover, appeal to explanatory considerations may in fact be important considerations for getting the Bayesian machinery up and running in the first place.

Probabilistic considerations therefore pose no threat to abductive inference, and thus this line of scepticism towards our ampliative inferences is similarly unsustainable.

2.4   Taking a stance

2.4.1   The empiricist dilemma

The final line of argument that van Fraassen offers in support of epistemic voluntarism is based upon the overall value of adopting such an epistemological outlook, the specifics of probabilistic coherence and ampliative inference aside. Essentially, the idea is that certain broadly meta-philosophical debates are not best understood within a traditional epistemological framework. To take the crucial example, van Fraassen's position within the philosophy of science – as is clearly implied by its name – is a modern-day empiricism about the natural sciences: it is thus premised upon the idea that our ultimate source of knowledge is experience; and it thereby entails a rejection of any attempt to know the nature of things beyond the limits of our senses. Crucially however, van Fraassen argues that empiricism cannot be taken as the adoption of a certain doctrine, and nor can it be characterised in terms of a specific *credo*: for van Fraassen, it is instead to be understood as a particular 'stance' that one adopts towards scientific inquiry, metaphysical speculation and any other putative source of knowledge. Tying this point together with the previous discussion, we can note that the positive and negative arguments for epistemic voluntarism sketch an epistemological picture whereby to make an epistemic judgement is to undertake certain commitments, and where our canons of inferential practice are to be considered permissible rather than obligatory. And so with empiricism in general: rather than the embracing of a doctrine or a set of core beliefs, empiricism is to be seen as the adoption of an epistemic policy or set of commitments; rather than adherence to a set of principles, it is a tendency to privilege experience, advocate scientific practice and to avoid metaphysical speculation. In particular, the 'empirical stance' can be especially well characterised in terms of one's views regarding explanation. According to van Fraassen, two central features of empiricism are the rejection of certain *demands* for explanation – and thus a willingness to stop the explanatory regress at an earlier stage than one's more metaphysically inclined opponent – and dissatisfaction with certain *types* of explanation, in particular those that proceed by the postulation of additional entities. Thus van Fraassen's (1977a; 1980: 97–157) pragmatic theory of explanation, which attempts to distinguish between good and bad explanations in terms of relevance to a context, rather than the invocation of laws of nature, natural kinds and other metaphysical villains.

There are a number of reasons for van Fraassen's insistence on this point. Firstly, he believes that it is misguided to conceive of the various episodes in the history of philosophy that we would call 'empiricist' in terms of a shared doctrine. What unites such otherwise disparate thinkers as the Oxford Nominalists, the British Empiricists and the Logical Positivists (to name but a few) is rather their shared opposition to the speculative metaphysics of their day. As van Fraassen sees it, 'the story of empiricism is a story of recurrent rebellion against a certain systematising and theorising tendency in philosophy: a recurrent rebellion against metaphysics' (van Fraassen, 2002: 36).

Secondly, and more importantly, characterising empiricism in terms of a particular doctrine actually threatens to undermine the position. The supposition that such a characterisation is available van Fraassen terms 'Principle Zero', the thought that

> For each philosophical position X there exists a statement X+ such that to have (or take) position X is to believe (or decide to believe) that X+.
>
> (2002: 41)

What we must then ask is what this statement X+ would be in the case of empiricism. Clearly, it would have to satisfy two basic desiderata – it would have to furnish the basis for the empiricist's so-called radical critique of metaphysics; and on pain of self-contradiction, it would also have to be immune to said radical critique. But these two requirements pull apart. The essence of the empiricist critique of metaphysics is that such speculation is done from the armchair and thus does not originate from, nor stand open to severe testing in the face of, the evidence of our senses. Consequently then, if our empiricist statement X+ is to be immune from its own radical critique of metaphysics it must be a *factual* thesis – based upon experience and open to empirical refutation. But if our empiricist statement X+ is a factual thesis, then disagreement with X+ must be admissible; a good empiricist may even say that such disagreement is in fact *desirable*, since another important aspect of the empiricist critique is that metaphysics leads to dogmatism. But if it is legitimate to disagree with our empiricist statement X+ (by denying that experience is the only source of knowledge, say), then such a statement can hardly furnish a radical critique; it would just be one opinion among many. Moreover, the empiricist would then be in the uncomfortable position of trying to *forbid* metaphysical speculation on the grounds that it runs counter to his central dictum that experience is our only source of knowledge, while at the same time trying to *endorse* disagreement with his central dictum on the grounds that all factual theses are open to debate. To put the same point another way, suppose we did encapsulate empiricism as, for example, the claim that experience is our only source of knowledge. What then would be the status of such a claim? It cannot be taken *a priori*, for then the empiricist would himself be embracing just the sort of speculative metaphysics that he supposedly rejects; he would be in the unfortunate position of reasoning from his armchair that one cannot in fact reason from one's armchair. Yet if the claim is taken *a posteriori*, as an empirical claim that itself needs to stand before the tribunal of experience, then the empiricist's doctrine becomes nothing more than a scientific conjecture, itself open to dispute and disagreement; crucially then, one could no longer maintain a 'radical critique of metaphysics', since empiricism would become something to debate with one's more metaphysically inclined opponents, rather than a 'recurrent rebellion' against their predilection for speculation (cf. van Fraassen, 2002: 43–44).[12]

Empiricism construed as a doctrine therefore leads to a straightforward contradiction: it either rejects speculative metaphysics by accepting a piece of speculative metaphysics or it advocates the policy of forbidding disagreement with its central dictum that experience is our only source of knowledge (the radical critique of metaphysics) with a policy that explicitly *endorses* disagreement with its central dictum on the grounds that it must be a factual thesis. The solution, according to van Fraassen, must therefore be to reject Principle Zero. There is no statement X+ which one must believe in order to be an empiricist; rather, the position consists of a certain attitude towards putative sources of knowledge, a certain view concerning the making of epistemic judgements and a certain policy towards our methods of reasoning. In short, empiricism can only be understood as a stance; and since – whatever else we may think about it – we don't take empiricism to be *incoherent*, the cogency of various meta-philosophical debates provides a general argument for epistemic voluntarism.

One can apply pressure to this argument from a number of directions. Jauernig (2007: 276–286) has argued that while van Fraassen's considerations show that certain formulations of a doctrinal empiricism are untenable, they do not show that *all* such formulations are untenable. The basic problem facing the doctrine empiricist is that he wishes to criticise the metaphysician for his disagreement with a particular factual thesis (that experience is our only source of knowledge) while maintaining the policy that disagreement with any factual thesis is admissible. The simple fix then is for the doctrine empiricist to take a slightly more self-reflexive policy on putative disagreements: rather than legitimising disagreement with *any* factual thesis, the doctrine empiricist need only legitimise disagreement with any factual thesis *other than the central empiricist dictum* (or indeed, anything logically entailed by that central empiricist dictum). The contradiction is then straightforwardly resolved, for the metaphysician's misguided views on the utility of armchair reasoning would be rendered inadmissible on the grounds of their deviation from the empiricist party line; while at the same time, such metaphysical speculation would not be rescued on the principle that any disagreement with a factual thesis is admissible, since the factual thesis they disagree with is not a specific empirical hypothesis framed by the natural sciences, but the core tenet of empiricism itself.

This at least is the basic strategy, but as Jauernig shows, it allows a more sophisticated implementation that seems somewhat less *ad-hoc*. Her preferred amendment to the empiricist policy concerning disagreements is as follows:

> any hypothesis that can in principle be empirically investigated is admissible as long as it has not been ruled out by the available empirical evidence, and only hypotheses that can in principle be empirically investigated are admissible.
>
> (Jauernig, 2007: 278)

The idea is that a commitment to empirical investigation is just as good a gloss on the vague idea that disagreement with any factual thesis is admissible, and arguably lends some important qualifications to such a notion. So Jauernig's proposal seems perfectly plausible. More importantly however, it also exemplifies the same sort of self-reflexivity as the proposal discussed above. Suppose that the empiricist's central dictum can be summarised as something along the lines of experience being our only source of knowledge, that this dictum is meant to provide a radical critique of metaphysics by rejecting any disagreement with this central claim about experience and that the empiricist is committed to the above policy about only admitting hypotheses open to empirical investigation. On pain of incoherence, let's also say that the empiricist's central dictum is itself open to empirical investigation, that it is a factual thesis. Still, the empiricist need not admit metaphysically loaded alternatives as admissible hypotheses: for while the advocacy of *a priori* reasoning may be acceptable insofar as disagreement with the privileging of experience is acceptable, such views will not themselves be open to empirical investigation and so will be ruled out by Jauernig's amended policy.[13]

Jauernig's claim then is that doctrine empiricism can be easily reformulated – in a principled, non-question-begging way – so as to avoid van Fraassen's dilemma, and hence to undermine the contention that empiricism can only be understood as the adoption of some kind of epistemic stance. The question however is not so much with the success of Jauernig's reformulation but with whether or not such a reformulation still offers a non-stance-based conception of empiricism (van Fraassen, 2004b: endnote c; 2007: 372–373). The worry is that Jauernig's basic strategy is to formulate an increasingly sophisticated policy for the empiricist to adopt with respect to determining admissible rival hypotheses, a policy that plays such a substantial role in the articulation and defence of her conception of empiricism that it is ultimately incompatibility with this *policy* – rather than with the core empiricist *doctrine* concerning the primacy of experience – that drives the supposedly radical critique of metaphysics. In short, it looks as if the doctrinal aspect of Jauernig's doctrine empiricism plays an entirely subservient role to the attitudes, commitments and policies that must now augment the position. And if this is the case, then rather than countering van Fraassen's argument for stance empiricism, Jauernig can in fact be seen to be simply re-iterating it.

The extent to which a purported amendment to doctrine empiricism is in fact a transition into stance empiricism is an excruciatingly subtle issue, and relatively inconclusive either way. But while we may have some sympathy for van Fraassen's contention that doctrine empiricism is untenable, a more damning criticism is that his argument may in fact prove too much. The basic problem facing the doctrine empiricist is that he wishes to hold certain beliefs that attempt to satisfy two competing desiderata – they must furnish the so-called radical critique of metaphysics by ruling out certain contrary theses concerning our acquisition of knowledge, and they must be immune to said radical critique by allowing (even encouraging) dissenting opinion. But as Ladyman (2004b: 139–140) points out – and as Mohler (2007) subsequently develops – the move to stance empiricism does not necessarily resolve this dilemma. For while it is true that one cannot characterise stance empiricism in terms of a particular doctrine or set of core beliefs, but rather in terms of various attitudes and epistemic commitments, it does not follow that the stance empiricist lacks such beliefs altogether: to believe that experience is our only source of knowledge may well not be *sufficient* to characterise the empirical stance, but it may still be *necessary*. And if the stance empiricist is committed to certain beliefs that attempt to furnish a radical critique of metaphysics while remaining immune from such a critique – albeit as part of a larger epistemological framework that deals ineliminably with attitudes and commitments – then he will face the same kinds of inconsistency as the doctrine empiricist will.[14]

Moreover, as Mohler (2007: 213–214) argues, it seems that no matter how the stance empiricist fleshes out the details of his position, he must be committed to exactly the sorts of beliefs that generate the problem at hand. Suppose that the empirical stance can be characterised in terms of a privileging of experience and a deep scepticism towards metaphysical speculation – it is undoubtedly more sophisticated than this, but this would be an appropriately coarse-grained starting point. Then although such a characterisation does not mention any central doctrine or core beliefs, it follows on pain of pragmatic incoherence that the stance empiricist must also believe something concerning the reliability of experience and the unreliability of armchair reasoning – something along the lines of experience being our only source of knowledge. The idea is that it would be straightforwardly incoherent to hold the stance empiricist's positive attitude towards empirical investigation, and negative attitude towards metaphysics, without holding corresponding beliefs *about* these putative sources of knowledge; for if he did, he would be in the situation of having to assert that he privileges experience as a source of knowledge even though he didn't *believe* it to be more reliable than the alternatives. The situation would be an instance of Moore's Paradox only concerning the stance empiricist's epistemic policy – 'my policy is to treat empirical investigation as more reliable than any other source of knowledge, but I do not believe empirical investigation to be any more reliable than any other source of knowledge.' The point of course generalises: however the stance empiricist's attitude, policy or commitments are to be characterised, there will also be a corresponding belief that he must hold on pain of the pragmatic incoherence of endorsing something he also denies believing. And since these beliefs will correspond to the fundamental orientation of stance empiricism, they will also all be the sorts of belief that will furnish the radical critique of metaphysics (experience is our only source of knowledge, disagreement with any factual hypothesis is admissible); consequently, these beliefs must all attempt to satisfy our diverging desiderata of furnishing the radical critique while remaining immune to said critique; and thus the original dilemma repeats itself.

In his original response to Ladyman, van Fraassen (2004a: 173) concedes that the stance empiricist will be committed to the sorts of beliefs that landed the doctrine empiricist in trouble, but that since these beliefs do not constitute the *basis* for his epistemic stance – we can think of them instead as simply the consequence of having a coherent view of one's own epistemic activity – they need not concern him in the same way that they concern his doctrinal colleague. In his response to Mohler, van Fraassen (2007: 374–375) elaborates on this idea in a little more detail. Clearly, the point cannot simply be that the stance empiricist acquires these problematic beliefs accidentally, or that these problematic beliefs play a less fundamental role in his conception of empiricism than they do for the doctrine empiricist, or that these problematic beliefs only concern the coherence of his epistemic activity rather than his characterisation of empiricism: as Mohler (2007: 215) stresses, that the stance empiricist holds such beliefs *at all* is sufficient for the argument to go through. The point rather is this: just as stance empiricism in general is not to be characterised in terms of a set of core beliefs, so is the stance empiricist's radical critique of metaphysics not to be characterised in terms of a set of core beliefs. The conflict between stance empiricists and their metaphysically inclined opponents lies not in their commitment to contrary beliefs, but in their commitment to contrary attitudes and policies concerning our putative sources of knowledge and explanatory practices. The stance empiricist's radical critique therefore should be seen in terms of holding a positive attitude towards empirical investigation, and a negative attitude towards dissenting attitudes towards empirical investigation. The upshot of all this is that since the supposedly problematic beliefs that the stance empiricist must hold – on pain of pragmatic incoherence – about experience being our only source of knowledge do not themselves play any role in the radical critique of metaphysics, it is simply irrelevant as to whether or not they meet the diverging desiderata that drove our dilemma. In short, since stance empiricism assigns no important role to beliefs *at all*, it doesn't really make any difference which beliefs he is required to hold.

2.4.2   Rationality and relativism

The preceding discussion shows that it is far from clear that construing empiricism as an epistemic stance offers any substantial advantages to construing it as a doctrine – and if one is willing to adopt an appropriately flexible understanding of what the core thesis of empiricism might be, it is far from clear that construing empiricism as an epistemic stance even offers any meaningful *alternative* to construing it as a doctrine. Yet we need not draw any definite conclusions regarding the relative merits of stance and doctrinal empiricism in order to assess this particular argument for epistemic voluntarism, since – whatever its comparative virtues – construing empiricism as an epistemic stance is an inherently undesirable option. There are two key respects in which this is the case: firstly, that regardless of the success or failure of its doctrinal ancestor, a stance-based empiricism is *utterly unable* to furnish the so-called radical critique of metaphysics that prompted van Fraassen's concerns with the nature of empiricism in the first place; and secondly, that the overall voluntarist picture of our epistemic lives – of which a stance-based empiricism is a manifestation – is too wildly divorced from our intuitive understanding of rationality to be credible. We should therefore reject van Fraassen's contention that the two basic strands of his voluntarist epistemology (to make an epistemic judgement is to undertake an epistemic commitment; rules of inference are permissible but not obligatory) together provide a superior understanding of certain broadly meta-philosophical debates: for not only does this combination fail to do justice to van Fraassen's *own* meta-philosophical posture, it also fails to connect with anything that we would even regard as a rational philosophical debate.

That stance empiricism cannot furnish a radical critique of metaphysics lies not in the fact that its basis for doing so is either not radical (since it is an empirical thesis open to dissent) or not a critique (since it is itself a piece of metaphysics), but in the fact that whatever force it has will be entirely context-relative. Thus much is at least suggested by van Fraassen's conclusion above that stances only compete with respect to their particular policies, rather in terms of their particular beliefs. More specifically, as Chakravartty (2004: 179–182; 2007: 191) argues, while an epistemic framework that demands no more of a philosophical position than its logical consistency and probabilistic coherence (and no more of its adherent than his appropriate commitment to the position) may well provide the empiricist room to articulate his own metaphysical disinclination without fear of self-contradiction, it also provides his opponent room to articulate his own metaphysical predilections without fear of compelling criticism. For if the only criteria in play for the adoption of one epistemic stance over another are the internal coherence of the stance and the individual values of the agent, then provided that the speculative metaphysician can demonstrate the logical consistency and probabilistic coherence of his own 'inflationary' epistemological policy (and of course, demonstrate his due commitment to the beliefs thereby acquired), there would appear to be no further standards by which to convict him. Of course, from the empiricist's point of view such an epistemological policy is quite wrongheaded: it asks explanatory questions for phenomena already well-understood, and seeks to explain them through the invocation of unobservable phenomena even more mysterious than that with which we began. But from the metaphysician's point of view, it is the empiricist's policy which is woefully inadequate: disavowing legitimate inferential methods and reaping an impoverished worldview as a consequence. Yet without further constraints beyond the internal coherence of their respective positions, and in the absence of any stance-transcendent values by which to adjudicate their conflicting epistemic goals, this seems to be as far as either side can proceed – not so much a radical critique as a refusal to play the same game.[15]

Of course, we shouldn't automatically assume that just because a disagreement concerns a set of competing values that no rational resolution is available. As van Fraassen (2002: 60–63; 2007: 375–378) has suggested, such resolutions are sought – and occasionally achieved – in the domains of ethics and politics; and rivals to the dispute may often have standards in common against which the relative merits of their conflicting values can be judged. But in the context of a radical critique of metaphysics, such hopes seem both overly optimistic and somewhat misleading. As Chakravartty (2004: 182) argues, those cases where one can propose a rational resolution of a competing set of values almost always proceed by way of a *reductio*: one shows how a particular ethical standpoint legitimises courses of action deemed abhorrent from the agent's own perspective, or how a particular policy of distribution disadvantages those it was supposed to benefit, and so on and so forth. But it is difficult to see how a similar situation could arise in the case of competing epistemic stances, which *ex hypothesi* are meant to be internally coherent. Certainly one cannot simply criticise each one of the metaphysician's inflationary views regarding explanation by postulation, metaphysically substantive laws of nature, instantiated universals and so on and so forth, since, as van Fraassen himself is at pains to point out, an epistemic stance cannot be identified with any particular set of beliefs – it is more like the policy by which one acquires one's particular set of beliefs. Moreover, even if one could demonstrate to the speculative metaphysician that *in general* his own epistemic commitments led him to conclusions in conflict with his particular views concerning, say, explanatory strength and abductive inference, one would not have shown the superiority of empiricism over metaphysics – one would merely have shown that the speculative metaphysician didn't even meet the minimal requirements for having a stance in the first place. Unless van Fraassen wishes to contend that the empirical stance is the only logically consistent and probabilistically coherent option on the market (which is clearly implausible), then he cannot hope to deliver a radical critique of one stance from the perspective of another, since one of the crucial pre-requisites for there being another stance to criticise in the first place is that it lacks the internal incoherence necessary for such criticism.

Stance empiricism cannot deliver the radical critique of metaphysics for which it is intended; and this perhaps is for the good, since as several commentators have maintained, metaphysical speculation may in fact be a necessary component of any satisfactory empiricism.[16] But we need not pursue this issue here – it is sufficient to note that van Fraassen's overall contention that the various strands of his minimalist conception of rationality together provide a more satisfactory framework for the understanding of various meta-philosophical debates *fails* since it cannot even accommodate his own conception of what constitutes his own philosophical position. Moreover, we might well question the desirability of such a framework in and of itself: for in addition to entailing epistemic relativism, and thereby rendering what we would intuitively regard as a substantial philosophical issue utterly irresolvable, it is in general difficult to see how the concept of rationality given to us by the various strands of van Fraassen's epistemology constitutes anything that we would intuitively regard as such. As Psillos (2007: 158–160) argues, it would seem that the voluntarist concept of rationality manages both to license *too much* – insofar as it counts as rational various doxastic combinations and epistemic policies that we would be loathe to consider as such – and to license *too little* – insofar as it counts as rational various doxastic deficits that we would similarly regard as absurd. That a voluntarist conception of rationality is too broad follows from the previous discussion: if the only constraint upon a legitimate epistemic stance is its internal coherence, then an awful lot of epistemological policies are going to count as rational. Indeed, as Ladyman (2004b: 141–142) notes, any method of ampliative inference can be made logically consistent and probabilistically coherent provided one is willing to 'update' the rest of one's web of belief appropriately – including inductive inference upon scant evidence, a refusal to make inductive inferences at all and even the perverse strategy of *counter-inducting* upon one's evidence (the more times it's happened in the past, the *less* likely it will happen in the future). But this can't be right: indeed, it is precisely on the basis of the blatant *irrationality* of the latter strategy that casinos make most of their money ('it's sure to come up black soon ...').

That voluntarism may license too little rests upon the fact that if our beliefs are rationally permissible rather than rationally obligated (at least, our beliefs concerning contingent, empirical states of affairs), then any epistemic agent reflecting upon his situation may well argue himself out of holding any beliefs at all: if the belief that p is rationally permissible then we have a *prima facie* case for disbelieving ¬p; and if the belief that ¬p is rationally permissible then we have a *prima facie* case for disbelieving p; any such agent may then simply refrain from holding any beliefs about any contingent matter of fact, and this cannot be part of our conception of rationality either (Psillos, 2007: 159–160). The whole point of epistemic voluntarism of course is that at certain junctures, it will simply come down to an agent's non-epistemic *values* over whether or not they believe that p or ¬p; the point though is that in order to do so, an epistemic agent must be conscious of the fact that whatever set of beliefs he has committed himself to have no greater epistemic credentials than any other set of beliefs he could have committed himself to – and this too seems deeply at odds with our understanding of a rational agent.

The general problem, as Psillos (2007: 156–158) diagnoses, is that the voluntarist conception of rationality is purely *structural*: rationality is held to reside in the (logical and probabilistic) *relationships* that hold between an agent's beliefs, with no regard to the *content* of any of those beliefs, and this 'goes against the deep-seated intuition that rationality has to do with what an agent does to make sure that her beliefs make contact with the world' (2007: 156). Basically, a purely structural conception of rationality has nothing to say about when one is to accept something as evidence, and so has no capacity to condemn as irrational an agent who is willing to ignore any information inconsistent with what they already believe. Consider the Flat Earther who, rather than attempting to explain why the Earth looks round from space, simply denies that it does so; or the Creationist who, rather than trying to accommodate the fossil record into their own historical account, simply denies that there is such a record to accommodate – both of these would be perfectly rational on the voluntarist account, which demands nothing more on an agent than that his beliefs fit together in the right way, regardless of what these beliefs may or may not be.[17]

There may be some room to manoeuvre here: we must be careful to distinguish between those cases where an agent is acting irrationally and those cases where he is (at least by our lights) simply mistaken. As van Fraassen notes, there are no objective standards concerning the relevance of a new piece of evidence; and thus whether or not an epistemic agent *should* accommodate any particular piece of evidence will be context-dependent. As he puts it,

> if the person in question has quite a different opinion of what it is that s/he knows or has received by way of evidence, we can say that s/he is mistaken, but not convict of irrationality on the basis that evidence did not receive its proper response.
>
> (van Fraassen, 2007: 354)

Again we can see voluntarism slipping into an unpalatable form of relativism: for just as it seemed intuitively philosophically undesirable to countenance an epistemological framework whereby all meta-philosophical debates are reduced to unassailable assertions of taste, so too does it seem intuitively philosophically undesirable to countenance an epistemological framework whereby one could always choose to disregard the evidence. Relevance may well be context-dependent; but as Psillos (2007: 157–158) maintains, there must be limits. To return to our example above: while we may be willing to accept some disagreement with the Flat Earther and the Creationist over the *importance* of certain phenomena (pictures from space, the fossil record), we would not be willing to accept their *denial* that such evidence even exists – and of course, Flat Earthers and Creationists take it as part and parcel of their views that such phenomena must be explained one way or another. But on the voluntarist story, such denials would be perfectly rational – if one believed that the Earth was created in 4004 BC, then the easiest way to maintain doxastic consistency in the face of a fossil record that appears to pre-date the act of creation would be to simply reject the notion that there could be such a fossil record altogether; and for the epistemic voluntarist, this would be just as acceptable as attempting to explain away such putative counterexamples. A concept of rationality that would legitimise the epistemological policy of never attending to any evidence cannot be *our* concept of rationality – the combination of van Fraassen's views regarding the commitments of an epistemic judgement and the inductive liberalism of our epistemic policies neither serves his own meta-philosophical purposes nor reflects anything that we would understand as rationality.

2.5 Summary

In this chapter, I have attempted to distinguish three distinct lines of argument that van Fraassen offers for epistemic voluntarism: a positive argument to the effect that independently compelling considerations over the diachronic probabilistic coherence of an agent's beliefs force us to reconceptualise the act of epistemic judgement as the undertaking of a particular epistemological commitment; a negative argument to the effect that any epistemological framework that attempts to add more substance to the epistemic voluntarist's minimal constraints of logical consistency and probabilistic coherence – in the way of rationally compelling rules of ampliative inference – is untenable; and the overall contention that in adopting the kind of epistemic framework entailed by the first two arguments, we are better able to make sense of various important meta-philosophical debates – and in particular, that which exists between the speculative metaphysician and the hard-nosed empiricist.

As I promised in Chapter 1, however, these last two arguments have struggled to maintain the delicate balance between advocating an epistemological framework that is much weaker than its traditional rivals and at the same time attempting to avoid an epistemological framework that is so weak that it simply collapses into scepticism and/or relativism. To argue for epistemic voluntarism is to argue for radical scepticism; and to endorse epistemic voluntarism is to endorse full-blown relativism. Neither conclusion is perhaps that surprising, in light of how we initially introduced van Fraassen's minimalist concept of rationality. We noted that the epistemic voluntarist – in contrast to the traditional epistemologist – denied that neither our current beliefs, nor our methods of belief revision, admitted of any independent justification; yet also denied – in contrast to the sceptic – that it was irrational to hold a belief, or follow a method of belief revision, that lacked such independent justification. More schematically, the epistemic voluntarist first endorses the sceptic's argument so as to undermine the traditional, rules-based epistemology that is his target, and then endorses the relativist's broader conception of rationality so as to avoid undermining his own alternative. Yet the epistemic voluntarist attempts to evade both a debilitating scepticism and an implausible relativism by only endorsing enough of the sceptic's argument to undermine a substantive epistemology (leaving his own minimalist epistemology intact), and only endorsing enough of the relativist's conception of rationality to license his own conception of an epistemic stance (without thereby licensing absolutely anything). Unfortunately, neither move is successful.

In the case of van Fraassen's critique of inference to the best explanation, we noted that unless one grants the general reliability of our methods of hypothesis ranking, such an argument would amount to nothing more than the sort of radically sceptical challenge that would equally undermine the view that abduction was rationally permissible as it would the view that abduction was rationally obligatory; yet in granting the reliability of our methods of hypothesis ranking, one can no longer doubt our methods of hypothesis generation against which van Fraassen's argument is directed. Similarly, in the case of van Fraassen's critique of the *a posteriori* justification of induction, we noted that unless one grants the fact that we need not know induction to be reliable in order for it to be reliable, and concedes that the mere possibility of error is irrelevant to the actual track record of our inductive practices, such an argument would be Humean scepticism all over again; yet in granting these two moves, our evidence for the occasional fallibility of our inductive practices was easily accommodated within the reliabilist framework. In both cases then, van Fraassen's negative argument for epistemic voluntarism can only undermine the traditional epistemological picture by undermining the prospects for the epistemological project altogether. And in the case of the overall utility of adopting a stance-based conception of our various meta-philosophical debates, we found that not only did such a move fail to capture the empiricist's so-called radical critique of metaphysics, but demanding nothing more of an epistemic policy that it be logically consistent in fact eliminated any possibility of meaningful comparison between differing philosophical positions altogether.

The strongest case made for epistemic voluntarism was therefore the first argument from diachronic probabilistic coherence. Yet while none of the considerations presented regarding the future-directed constraints upon an agent's distribution of subjective probabilities led us into grievous error, there remained a serious argumentative lacuna. In short, the constraint of diachronic probabilistic coherence was only compelling insofar as we had *already* adopted a voluntarist perspective: diachronic Dutch-books indicate an incoherence between sets of beliefs that only a voluntarist would suppose needed to be coherent; and to worry, for each set of beliefs, how we may come to revise those beliefs is to endorse a notion of epistemic integrity that presupposes the future-orientated concerns of the epistemic voluntarist. Diachronic probabilistic coherence and epistemic voluntarism are part of the same package; thus neither can be used to provide an argument for the other.

Thus there are no good arguments for epistemic voluntarism; yet provided one is willing to swallow a degree of epistemic anarchy at the meta-philosophical level, there is nothing straightforwardly inconsistent with being an epistemic voluntarist nonetheless. However, we have already seen in Chapter 1 that epistemic voluntarism is an unnecessary component to the articulation and defence of constructive empiricism, insofar as those arguments that it seeks to undermine are not terribly convincing arguments anyway. And in Chapter 3, we shall encounter those arguments that epistemic voluntarism cannot resolve, epistemic anarchy notwithstanding.

3
Against Epistemic Voluntarism: Musgrave, Modality and Mathematics

3.1 The problem(s) of internal coherence

At its heart, constructive empiricism consists of a fundamental distinction between the claims that our accepted scientific theories make regarding observable and unobservable phenomena; and this in turn divides criticisms of the position into two broad categories. The first, discussed in detail in Chapter 1, concerns the epistemic relevance of such a distinction: and we have seen how van Fraassen's voluntarist framework provides the resources for resolving – or perhaps better, dissolving – objections of this kind (even if we have subsequently seen reasons to doubt the attractiveness of such an epistemological perspective in Chapter 2). The second category of criticisms concerns the internal coherence of constructive empiricism, that is, with whether or not the constructive empiricist can consistently maintain his own distinction between the observable and the unobservable. This second category of criticism is to my mind the more serious, for not only does it charge constructive empiricism with being *untenable* (rather than merely *unattractive*), it is also clear that the epistemic relevance of the constructive empiricist's distinction can only be brought into question once it has been established that such a distinction can be coherently drawn in the first place.

This chapter will be concerned with this issue of internal coherence. As has already been briefly mentioned in Chapter 1, the general problem facing the constructive empiricist is that in attempting to specify the limits to what he believes, he must inevitably transcend those very limitations. This is illustrated with particular clarity in Alan Musgrave's (1985) famous complaint that in order to specify the distinction between observable and unobservable phenomena, the constructive empiricist must in fact believe certain claims about the unobservable phenomena (namely, that these entities and processes are indeed unobservable), thus violating the epistemic policy of only believing those claims that are about observable phenomena that such a distinction was meant to establish. The exact same dialectical structure however – that in order to draw a line, one must first step beyond it – can also be seen to underlie a number of other challenges to the internal coherence of constructive empiricism, in particular James Ladyman's (2000) contention that the constructive empiricist cannot provide a satisfactory account of modality, and Gideon Rosen's (1994) argument to the effect that constructive empiricism is incompatible with the sort of mathematical nominalism that it seems to entail. My first aim in this chapter therefore is the purely taxonomic one of showing how these seemingly disparate debates are structurally equivalent to the basic objection first posed by Musgrave, and how this then helps us to understand their subsequent argumentative manoeuvres.

What this imposition of structure ends up revealing however is the ineliminable role that van Fraassen's epistemic voluntarism also plays in his response to these kinds of criticisms. In all three cases, the constructive empiricist faces a dilemma between believing *too much* – the unobservability of unobservable phenomena; the objective truth-conditions of counterfactual conditionals; the existence of abstract mathematical objects – and thereby violating the epistemic policy that his view regarding the aim of science was meant to establish, and believing *too little*, and ending up with a hopelessly impoverished philosophical position. And in all three cases, van Fraassen's considered strategy is to hedge his bets on the side of believing too little, while attempting to make up the doxastic difference with various amendments and fixes. As we will see however, the problem with this strategy is that many of these patches possess little justification beyond their ability to finesse the difficulty at hand, and many seem straightforwardly to beg the question. In short, van Fraassen's responses to these various objections only pass epistemic muster within a context that explicitly denies that there is anything more to providing a satisfactory resolution beyond meeting the basic constraints of logical consistency and probabilistic coherence. Thus *all* aspects of van Fraassen's articulation and defence of constructive empiricism rely essentially upon the adoption of his minimalist conception of rationality; yet while such a framework may have some merit with respect to questions of epistemic relevance (and we have seen reason to doubt this too), such systematic evasion of what we might call the hard question of the internal coherence of one's position can only be seen as a serious failing.

Of course, merely pointing out how van Fraassen's thoroughgoing epistemic voluntarism seems intuitively unsatisfactory (at least to those who have not already adopted such a perspective) will not significantly advance our criticism of constructive empiricism beyond the various concerns raised in Chapter 2, even if we agree that in moving to the question of internal coherence we thereby raise the stakes to what was at issue in our consideration of the question of epistemic relevance. However, in pursuing these related difficulties of unobservability, modality and mathematics we can finally unearth a serious tension in van Fraassen's articulation and defence of constructive empiricism that even the epistemic voluntarist must concede. For in addition to sharing a similar structure and developing dialectic, our trio of objections are also related to one another in terms of their increasing levels of abstraction. As I shall argue, van Fraassen's considered response to Musgrave's objection (a response developed in collaboration with Fred Muller) ultimately rests upon making various stipulations about the modal scope of certain beliefs held by the constructive empiricist. Consequently, the strength of Muller and van Fraassen's voluntarist response must rest in part upon the coherence of their philosophical account of modality; their response – assuming that one is willing to grant their voluntarist framework – effectively shifts the burden of proof into the domain of Ladyman's objection. And similarly, van Fraassen's considered response to the problem of modality (this time developed in collaboration with Bradley Monton) makes an ineliminable appeal to a model-theoretic account of modality, thus raising the question over the constructive empiricist's view of abstract mathematical objects and thereby shifting the problem – again, assuming that one is willing to grant their voluntarist framework – to the concern raised by Rosen. Crucially, however, once we ascend to this level of abstraction, the epistemic voluntarist strategy that we see employed with respect to Musgrave and Ladyman finally runs out of steam. For the doxastic deficit that the constructive empiricist must make up in taking the deflationary horn of Rosen's dilemma concerns the notion of logical consistency itself. Consequently, the constructive empiricist cannot justify his response to the problem of mathematics by appealing to a minimalist conception of rationality that demands nothing more from its advocates than logical consistency and probabilistic coherence, for it is those very notions that the constructive empiricist's response is attempting to secure. It is at these dizzying heights then that the voluntarist strategy finally unravels – for if the problem of mathematics remains insoluble (within a voluntarist framework), then there is no well-established model-theoretic account of modality for the constructive empiricist to appeal to in his response to Ladyman; if there is no well-established model-theoretic account of modality (within the voluntarist framework), then there is no deflationary account of modality for the constructive empiricist to appeal to in response to Musgrave; and if there is no deflationary account of modality (within the voluntarist framework), then Muller and van Fraassen's original solution is rendered promissory at best.

From the beginning, the overriding suspicion aroused by van Fraassen's minimalist epistemology was that it was too insubstantial to do any philosophical heavy-lifting. We have seen how a voluntarist framework can offer a kind of resolution for the various worries concerning the epistemic relevance of the constructive empiricist's central distinction between observable and unobservable phenomena; yet in Chapter 2 we have also seen that there is little to motivate a voluntarist framework beyond its ability to resolve such worries. In this chapter I show how epistemic voluntarism is simply unable to resolve the more pressing concern over the internal coherence of the constructive empiricist's central distinction. The conclusion is clear: the constructive empiricist should not be an epistemic voluntarist.

3.2   Musgrave's objection revisited

3.2.1   The unobservability of unobservables

Let us begin then with what I take to be the purest form of the problem facing the internal coherence of constructive empiricism. According to the constructive empiricist (as we have seen), what counts as observable-for-us is to be measured with respect to the various epistemic limitations of human beings considered *qua* measuring devices: contingent and empirical limitations that we can assume 'will be described in detail in the final physics and biology' (van Fraassen, 1980: 17). What this means then is that in order for the constructive empiricist to determine whether or not a scientific theory is empirically adequate, he must first of all appeal to our best scientific theories concerning the behaviour of light and the intricacies of human physiology to tell him which consequences of that scientific theory are about observable phenomena. To put the same point more succinctly: in order to draw his crucial distinction between observable and unobservable phenomena, the constructive empiricist needs first of all to appeal to his best scientific theory of observability – or strictly speaking, some suitable conjunction of his best scientific theories of the propagation of light, the resolution of the human eye and so forth – to tell him the identity of the observable phenomena.

And all of this raises an interesting difficulty. Constructive empiricism is of course the view that 'science aims to give us theories that are empirically adequate; and acceptance of a theory involves as belief only that it is empirically adequate' (van Fraassen, 1980: 12). Thus when the constructive empiricist accepts a successful scientific theory, he is committed to the view that he need only believe those consequences of the theory regarding observable phenomena in order to make sense of scientific practice; if he believed more than what his accepted scientific theory claimed about observable phenomena – or more importantly for what follows, if he *had* to believe more than what his accepted scientific theory claimed about observable phenomena – then constructive empiricism would be straightforwardly refuted as a thesis regarding the aim of science. As a particular case in point then, when the constructive empiricist accepts the conjunction of his various scientific theories of observability – call this conjunction the constructive empiricist's theory of observability T^* – he is committed to the view that he need only believe those consequences of T^* that concern observable phenomena. It follows then that, in order to know which consequences of T^* he need only believe, the constructive empiricist needs to know which consequences of T^* are about observable phenomena. However, it is T^* itself that tells the constructive empiricist what counts as an observable phenomenon: the constructive empiricist therefore needs to appeal to T^* in order to know which consequences of T^* he can believe.

Of course, the mere fact that the distinction drawn by T^* must also apply to itself is not an immediate cause for alarm. One asks the constructive empiricist which parts of his theory of observability he need only believe, and he responds with a rigorous criterion; it just so happens that this criterion is to be supplied by the theory of observability in question. There is certainly a sense then in which the constructive empiricist's distinction between observable and unobservable phenomena is circular, but not all circles are vicious: some circles are perfectly benign, and some even positively support their own application. To take a familiar example, Popper's well-known distinction between what he called science and non-science was based on the criterion of falsifiability, a criterion that as a simple matter of consistency must also apply to itself. Yet Popper's distinction – whatever else one may think of it – remains perfectly intelligible: even though his distinction between science and non-science rendered the criterion itself *non-scientific* (the criterion of falsifiability is not itself falsifiable), it did not render itself *incoherent.*

Unfortunately, it is far from clear that the constructive empiricist's position is similarly inoculated. The basic problem facing the constructive empiricist is that he has no guarantee that the parts of his theory of observability that he *needs* to believe in order to draw his distinction between observable and unobservable phenomena will coincide with those parts of his theory of observability that he *can* believe if his view regarding the aim of science is correct. The constructive empiricist is committed to the view that he need only believe those consequences of his accepted scientific theories that are about observable phenomena; thus he is committed to the view that he need only believe those consequences of his theory of observability that are about observable phenomena. A serious problem seems to arise then if those crucially important consequences of the constructive empiricist's theory of observability that tell him what *counts* as an observable phenomenon are not themselves consequences of the constructive empiricist's theory of observability that are *about* observable phenomena. Or, in other words, if those parts of his theory of observability that tell the constructive empiricist what he can believe are not *themselves* parts of his theory of observability that he can believe, then he is in serious trouble.

This is the concern raised by Alan Musgrave (1985), who argues that in fact some of the claims that the constructive empiricist needs to believe are not themselves claims that he can believe. For some of the consequences of his theory of observability that the constructive empiricist needs to believe in order to draw a principled distinction between observable and unobservable phenomena (and thus in order to determine whether or not a scientific theory is empirically adequate) will be of the form 'x is unobservable', which is clearly *not* about an observable phenomenon – and thus presumably not the sort of consequence of his theory of observability that the constructive empiricist can believe, insofar as he takes the aim of science to involve no more belief than that his accepted scientific theories are empirically adequate. The constructive empiricist's distinction between observable and unobservable phenomena is therefore not so much Popperian as Positivist, for just as the verification principle specified a criterion for meaningfulness that it itself apparently failed to meet, so the constructive empiricist's distinction between the observable and the unobservable specifies a criterion for belief that it itself fails to meet. Consequently, since the distinction between observable and unobservable phenomena is fundamental to constructive empiricism, and since this is a distinction the specification of which the constructive empiricist himself cannot believe, Musgrave (1985: 208) concludes that the position is 'untenable'.

3.2.2 On the syntax and semantics of observability

By now, van Fraassen's official response to Musgrave's objection has become somewhat notorious, and has admitted of several different interpretations.[1] The response is worth quoting at length:

> Musgrave says that '[x] is not observable by humans' is not a statement about what is observable by humans. Hence, if a theory entails it, and I believe the theory to be empirically adequate, it does not follow that I believe that [x] is not observable. The problem may only lie in the way I sometimes give rough and intuitive rephrasings of the concept of empirical adequacy. Suppose [T*] entails that statement. Then [T*] has no model in which [x] occurs among the empirical substructures. Hence, if [x] is real and observable, not all observable phenomena fit into a model of [T*] in the right way, and then [T*] is not empirically adequate. So, if I believe [T*] to be empirically adequate, then I also believe that [x] is unobservable if it is real. I think that is enough.
>
> (van Fraassen, 1985: 256)

In more recent work, however, Muller and van Fraassen (2008) have made the strategy a little clearer. They claim (2008: 200) that Musgrave's objection presupposes a *syntactic* account of theories (the view that a scientific theory consists of a set of propositions or other linguistic entities) – as exemplified by the demand that the constructive empiricist give us some account of how he can believe a consequence of his theory of observability of the form 'x is unobservable' – whereas constructive empiricism has always been explicitly wedded to a *semantic* account of theories (the view that a scientific theory consists of a set of models); and that once this is appreciated, one can see that the constructive empiricist need not violate his position with respect to the aim of science in order to draw his fundamental distinction, since to believe that an entity or process is unobservable is simply to believe that his empirically adequate theory of observability fails to classify that entity or process as observable (Muller and van Fraassen, 2008: 200–201).

The first claim is false, since both Musgrave's objection and, indeed, the response offered by Muller and van Fraassen are perfectly neutral between the syntactic and semantic approaches to scientific theories.

The exact same objection can be put in explicitly model-theoretic terms. The constructive empiricist – if he is correct in his view regarding the aim of science – only needs to believe those parts of his accepted scientific theories (i.e., those substructures of the models that constitute his accepted scientific theories) that are about (i.e., embed the representations of) observable phenomena. Thus in order to know which parts of his accepted scientific theories to believe, the constructive empiricist needs to appeal to his best scientific theory of observability (again, call it T*) to tell him the identity of the observable phenomena. It follows then that the constructive empiricist is committed to the view that he only needs to believe the empirical substructure of T* (i.e., the part of T* that embeds the representations of the observable phenomena) to be an accurate representation of the world. Thus, if T* classifies a particular entity or process as unobservable – that is, if the representation of that entity or process is not embedded within the empirical substructure of T* – then that is a representation of T* that the constructive empiricist cannot believe to be accurate, insofar as he takes the aim of science to involve no more belief than that his accepted scientific theories are empirically adequate. So just as the constructive empiricist cannot believe a consequence of his theory of observability of the form 'x is unobservable' since it is not a consequence of his theory of observability that is about an observable phenomenon, he also cannot believe a representation of x as an unobservable phenomenon since this would not be a representation that was embedded within the empirical substructure of the model. Whichever way one chooses to put it then – whether syntactically or semantically – Musgrave's objection remains the same: in attempting to draw a line between what he can and cannot believe, the constructive empiricist finds that he must first step beyond it.

The problem then may not simply lie with van Fraassen's penchant for 'rough and intuitive rephrasings'. Muller and van Fraassen's second claim however – that beliefs about unobservable phenomena are somehow reducible to (exhaustive) beliefs about observable phenomena – is more interesting, not least because pursuing this idea will quickly bring us into contact with van Fraassen's epistemic voluntarism.

According to Muller and van Fraassen then, to believe that (according to our best accepted scientific theory) a particular entity or process is unobservable does not commit the constructive empiricist to believing more than what his scientific theories say regarding the observable phenomena, since such a belief is in fact reducible to various perfectly legitimate beliefs regarding observable phenomena *along with* the belief that his theory of observability is empirically adequate. To take a slight variation on their example, suppose that the constructive empiricist accepts a scientific theory that classifies entities as either being an electron or being observable; and suppose further that there is no model within the set of models that constitute the theory in question in which these two categories overlap. It follows from this that if the constructive empiricist believes this theory to be empirically adequate – and therefore believes that all of the actual observable phenomena are represented as such in some model of that theory – then since he also knows that nothing that is classified as being observable by any model of that theory is also classified as being an electron by any model of that theory, then he must also believe that there are no entities that are both electrons and observable. In other words, he must also believe that there are no observable electrons, which is just to say that electrons are unobservable.

Another way to see what is going on here is to note that in his original complaint, Musgrave only objects to the constructive empiricist believing those consequences of his accepted scientific theories that are about unobservable phenomena; those consequences of his accepted scientific theories that are about observable phenomena however appear by contrast to be just the sort of thing that the constructive empiricist *can* believe. This then suggests a rather straightforward response to Musgrave's objection, for it looks as if the constructive empiricist can draw his crucial distinction between the observable and unobservable phenomena by simply reading off all of the observable phenomena from his theory of observability. In other words, since the constructive empiricist can provide necessary and sufficient criteria for being an observable phenomena – being classified as such by his theory of observability – and since presumably anything which isn't an observable phenomenon must therefore be an unobservable phenomenon, then it looks as though the constructive empiricist can draw his distinction between the observable and the unobservable after all, entirely from the observable side of the line.

This is what van Fraassen means when he argues that, if a theory has a consequence of the form 'x is unobservable', then it is simply the case that the theory in question does not have a model which classifies x as being an observable phenomenon ('...has no model in which [x] occurs among the empirical substructures'). In effect then, van Fraassen's response to Musgrave is to argue that the constructive empiricist does not need to believe what his accepted scientific theories say about the unobservability of unobservable phenomena in order to draw his fundamental distinction, since to believe that x is unobservable is just to believe that his empirically adequate theory of observability does not classify x as observable ('...if I believe [T*] to be empirically adequate, then I also believe that [x] is unobservable if it is real'). The same strategy of course has an obvious parallel in terms of syntax: since any consequence of an accepted scientific theory involving the predicate 'unobservable' is clearly problematic, the constructive empiricist should restrict himself to only believing those consequences of his accepted scientific theories that involve the predicate 'observable'; then, since any entity or process not predicated as observable is presumably unobservable, and since the constructive empiricist believes his theory of observability to be empirically adequate, he can infer that anything not predicated as observable by that theory must in fact be an unobservable phenomenon.

3.2.3   A question of scope

The response to Musgrave outlined above definitely has some merits but, as Muller and van Fraassen (2008: 201–204) themselves note, it can only get the constructive empiricist so far. To believe that one's theory of observability is empirically adequate is to believe that one's theory of observability correctly identifies all of the *actual* observable phenomena; according to Muller and van Fraassen's account then, to believe that electrons are unobservable is to believe that there are no observable electrons among the *actual* observable phenomena. But the belief that electrons are unobservable presumably goes beyond this: it is to believe a certain modal claim about the unobservability of electrons – that all possible electrons are unobservable, not just the ones that might actually exist. There are many ways to illustrate this point, the most obvious being to note that the unobservability of electrons is a law-like statement and therefore should support various counterfactual conditionals about what would happen if there were more electrons than there actually are; yet on Muller and van Fraassen's account, the belief that electrons are unobservable tells us absolutely nothing about whether or not we would be able to observe an additional electron, were it to exist. Thus in a pattern that we will see repeated throughout this chapter, Muller and van Fraassen's attempt to secure the coherence of constructive empiricism in terms of a suitably parsimonious set of belief falls woefully short of its intended target.

In order to bridge this residual gap, Muller and van Fraassen (2008: 204) propose an amendment to the constructive empiricist's basic epistemic policy: they stipulate that when it comes to matters of observability, exhaustive beliefs concerning the empirical adequacy of one's theory of observability are automatically to be given the widest possible modal scope: if the constructive empiricist's theory of observability classifies an entity or process as (actually) unobservable – that is to say, if the constructive empiricist's theory of observability fails to classify an entity or process as (actually) observable – then he is simply to believe that entity or process to be (necessarily) unobservable. The sheer awkwardness of this policy with respect to the constructive empiricist's theory of observability can be seen when we reflect on the fact that it presumably does not hold for any other of our classificatory categories. None of the scientific theories that I believe to be empirically adequate – that is, none of the scientific theories that I believe to have correctly classified all of the actual observable phenomena – classify any entity as being a sphere with a diameter greater than 10 miles, and as being made entirely out of gold. I therefore believe that there are no actual golden spheres with a diameter greater than 10 miles; I don't however believe this combination to be *impossible* in the same way that I discount the possibility of an *uranium* sphere with a diameter greater than 10 miles, or indeed in the same way that I discount the possibility of an observable electron. And nor should I. Muller and van Fraassen's considered response to Musgrave's objection therefore rests upon assigning a privileged status to what the constructive empiricist's accepted scientific theories say about observability which, while not necessarily *ad hoc*, certainly lacks any independent motivation: observability, although obviously of enormous importance to the constructive empiricist, is in all other respects a perfectly straightforward scientific concept, to be investigated and determined by our accepted scientific theories like any other scientific concept.

It is at this point then that the appeal to van Fraassen's voluntarist epistemology becomes paramount. Musgrave's objection challenges the constructive empiricist to draw a principle distinction between observable and unobservable phenomena, given that the various beliefs about the unobservability of unobservable phenomena that drawing such a distinction requires are in fact rendered illegitimate by the very position that such a distinction was meant to establish. Muller and van Fraassen's response is to argue that such putatively problematic beliefs are in fact completely reducible to more legitimate beliefs – provided of course that we are willing to grant this reduction class a privileged modal status that has absolutely no other motivation than to avoid Musgrave's objection. Within a traditional – that is, non-voluntarist – epistemological framework, one might reasonably query exactly how it is that the constructive empiricist can simply *stipulate* that his epistemic policy is sufficient to recover the beliefs necessary for his fundamental distinction between the observable and the unobservable, when the objection facing him is precisely that his epistemic policy is not sufficient to recover such beliefs. In other words, given that Musgrave has challenged the constructive empiricist's ability to consistently maintain his own distinction between observable and unobservable phenomena, there is something epistemologically very unsatisfactory in being told by Muller and van Fraassen that the constructive empiricist *can* maintain such a distinction on no stronger grounds than that if he is to be a constructive empiricist, he *must* maintain such a distinction.

The only way in which Muller and van Fraassen's considered response to Musgrave could be considered as anywhere near satisfactory would be if we were to adopt the sort of voluntarist framework of which van Fraassen is so enamoured, and which explicitly rejects any substantive epistemology beyond the basic constraints of logical consistency and probabilistic coherence. For while Muller and van Fraassen's amended epistemic policy may strike us as straightforwardly begging the question against Musgrave, it must at least be conceded that their strategy is perfectly rational according to this more parsimonious perspective. Indeed, in this respect Muller and van Fraassen's strategy for defending the constructive empiricist's ability to draw his central distinction exactly parallels the general voluntarist strategy for defending the epistemic relevance of the constructive empiricist's distinction: in the latter case, although we may not be able to see why claims about observables are any better warranted than claims about unobservables, we must at least concede that there is nothing inconsistent with maintaining such a distinction; in the former case, although we may not be able to justify Muller and van Fraassen's stipulative response to Musgrave, we must similarly concede that there is nothing incoherent in its formulation.

3.3   Ladyman's dilemma: Counterfactuals, contexts and conventions

3.3.1   The problem of modality

Musgrave's objection consisted of the claim that the constructive empiricist cannot consistently maintain his central distinction between observable and unobservable phenomena without thereby violating the explicit view regarding the aim of science that such a distinction was meant to establish; and we have seen how the official response to this problem makes an ineliminable appeal to the minimalist epistemological framework associated with van Fraassen's epistemic voluntarism. More specifically, we have seen how Muller and van Fraassen's considered response to Musgrave makes an ineliminable appeal to van Fraassen's minimalist epistemology in order to justify – or perhaps better, finesse any need to justify – various stipulations that they wish to make concerning the modal scope of the constructive empiricist's beliefs about observability. In addition then to our mounting worries over the philosophical adequacy of such a strategy, this response also raises the issue of the constructive empiricist's theory of modality in general: questions of justification or voluntarism aside, if the constructive empiricist is going to attempt to respond to Musgrave by way of a series of stipulations concerning the modal scope of his various beliefs, there had better be a satisfactory theory of modality to back it all up. This in turn however has also been challenged; and what is of particular interest for our present purposes is that the objections made to the constructive empiricist's theory of modality, and the responses made on its behalf, exemplify exactly the same dialectic structure that was found to underlie the debate surrounding Musgrave's original objection.

The problem of modality, as originally raised by James Ladyman (2000; see also Ladyman and Ross, 2007: 107–111), is that in drawing a principled distinction between observable and unobservable phenomena, the constructive empiricist is thereby immediately committed to various counterfactual consequences of his accepted scientific theories – what we would have observed, had the circumstances been different – of which he is fundamentally unable to provide a satisfactory analysis; and this is due to what appears to be a basic tension between van Fraassen's articulation and defence of constructive empiricism and his broader hostility towards metaphysical speculation. More specifically, Ladyman (2000: 849–852) argues that since van Fraassen holds a radically deflationary account of modality – to be expanded upon shortly, but which in essence holds that our modal claims are made true or false depending upon which models of our accepted scientific theories we choose to privilege – the distinction between observable and unobservable phenomena that is so central to constructive empiricism is rendered 'entirely arbitrary' (Ladyman, 2000: 850). Consequently, Ladyman argues, since the constructive empiricist's distinction between the observable and the unobservable must aim to capture some objective feature of reality if it is to do any kind of philosophical heavy-lifting, van Fraassen faces the unfortunate dilemma of either abandoning his deflationary metaphysics or abandoning constructive empiricism altogether.

Moreover, this initial tension can in fact be further developed to provide a general dilemma for any attempted defence of the modal dimensions of constructive empiricism, with obvious parallels to the objection raised by Musgrave in the previous section. As we have seen in the discussion of some of Churchland's objections in Chapter 1, in order to draw a principled distinction between observable and unobservable phenomena the constructive empiricist is committed to believing in some of the modal consequences of his accepted scientific theories: after all, there are some entities and processes that will never actually be observed, yet which the constructive empiricist would still presumably wish to classify as observable – the various moons around a distant planet, say, or the striped patterns upon some long-deceased dinosaur. The question then for the constructive empiricist is whether or not the observability of a particular entity or process (i.e., its being observable-in-principle) is itself an objective modal fact. For on the one hand, if the constructive empiricist maintains that these counterfactual conditionals do have objective truth-conditions (in terms of, say, the existence of other possible worlds), then he is admitting that in order to defend his view that the aim of science is merely empirical adequacy, he must in fact believe his accepted scientific theories to be more than empirically adequate in that he must believe them to correctly describe certain *non-actual* phenomena; on the other hand, if the constructive empiricist defends a non-objectivist account of his counterfactual truth-conditions (in terms of, say, certain linguistic conventions), he appears to be left with far too arbitrary a distinction with which to do any useful philosophical work. The constructive empiricist ends up believing either *so much* that he undermines his position or *so little* that he is unable to establish it in the first place.

3.3.2 From modal realism to structural realism

The most perspicacious way in which to understand Ladyman's dilemma is to see it in essence as a problem concerning the similarity ordering of possible worlds, which can be seen by taking the two horns of the dilemma in turn. Initially, one may think that the first horn of the dilemma is primarily an epistemological problem – that if there are objective modal facts about the observability of certain phenomena, then it is inconsistent for the constructive empiricist to claim to be able to know such facts since this would be to know certain facts about *unactualised possibilities*, and such knowledge would seem to go far beyond the limits set by the constructive empiricist's epistemic policy. Indeed, in his initial presentation of the problem, this appears to be Ladyman's view of the matter:

> It would be bizarre to suggest that we do not know about electrons merely because they are unobservable, but that we do know about non-actual possibilia. If we were to believe what our best theories say about modal matters, then why not believe what they say unobservables, too?
>
> (2000: 855)

This initial line of thought however is mistaken, as Ladyman himself is quick to note (2004a: 763–764). For van Fraassen, constructive empiricism is part of an overall empiricist package; he motivates the position (at least in part) in terms of being the best way to understand scientific practice, given that he wants to be an empiricist. As such, *van Fraassen* is indeed committed to the rejection of some kind of modal realism or modal objectivism, since that is part of his philosophical worldview. As Ladyman (2004a) points out, van Fraassen (2002) not only frequently characterises empiricism as a sceptical stance towards metaphysics (as discussed at length in Chapter 2), but in some cases even seems to suggest that the denial of objective modality is a *definitive* component of empiricism (e.g., van Fraassen, 1977b). But one can be a constructive empiricist without sharing van Fraassen's overall vision. As I have taken pains to stress, in its original formulation constructive empiricism is merely a view regarding the aim of science, to be situated within a broadly voluntarist epistemology (at least according to van Fraassen). As such, it is consistent with being a constructive empiricist to take any epistemological attitude one chooses towards any type of entity or process one chooses, insofar of course as that attitude is logically consistent and probabilistically coherent: one can be a constructive empiricist and believe in the existence of electrons; and in particular, one can be a constructive empiricist and believe in the existence of other possible worlds. Monton and van Fraassen make just this point in response to Ladyman:

> roughly speaking, a theory is empirically adequate iff what it says about the observable things in this world is true. Thus characterised, constructive empiricism is neutral on the issue of modal realism – unlike, to be sure, van Fraassen's overall conception of empiricism.
>
> (2003: 406)

Indeed, when considered in isolation, it is very difficult to see why believing something about merely possible phenomena should force anyone – be they constructive empiricist or otherwise – into believing something about actual unobservable phenomena. For although both types of phenomena under discussion share the characteristic of being unobserved, one must concede that claims made about *non-actual* phenomena are importantly different from claims made about *actual* phenomena (observable or otherwise). In particular, our epistemic access to merely possible entities or processes is completely different from our epistemic access to actual unobservable entities or processes. In the latter case, one arguably has some kind of causal interaction with the phenomenon in question (assuming of course that such phenomena exist), and our knowledge of such phenomena (assuming of course that we have such knowledge) is to be inferred from such interactions: we may, for example, infer the charge on an electron – and indeed, the existence of electrons in general – as the best explanation for the various occurrences that we have observed in a cloud chamber. By contrast, our knowledge of merely possible entities or processes derives from – to take the most prominent proposal as an example – some kind of *principle of recombination*, whereby distinct parts of the actual world can be patched together to create non-actual but possible worlds.[2] In other words, our knowledge of non-actual phenomena is based upon the unrestricted re-arrangement of our knowledge of actual phenomena; or as Lewis puts it, 'the principle ... that anything can coexist with anything else, at least provided they occupy distinct spatio-temporal positions. Likewise, anything can fail to coexist with anything else' (1986: 88).[3]

Our knowledge of merely possible phenomena and our knowledge of actual unobservable phenomena, therefore, depend upon quite distinct epistemological principles: recombination and inference to the best explanation, respectively. More importantly, these are epistemological principles towards which one can take very different attitudes: we have seen of course just how outspoken a critic of inference to the best explanation van Fraassen is; but there is no reason to suppose that he must therefore extend this hostility towards the principle of recombination invoked by the modal realist. The constructive empiricist can therefore reject knowledge of the actual unobservable phenomena without thereby committing himself to any particular epistemological attitude towards the merely possible.

Nevertheless, Muller (2005: 70) has objected that although we must concede that constructive empiricism and modal realism are *logically* compatible – after all, one is a view about the aim of scientific practice, the other a view about the nature of modality – our initial epistemologi-cal concern was well-motivated insofar as the two positions are in some sense *philosophically* incompatible. Muller argues that the pivotal moti-vation for constructive empiricism is to make sense of scientific practice without recourse to any kind of inflationary metaphysics; consequently, to combine constructive empiricism with modal realism is to scandalise the entire constructive empiricist enterprise. But as we should be aware by now, this is simply misguided. There are many motivations for con-structive empiricism, and there is no reason to take anti-metaphysical sentiments as the most important. Indeed, not even van Fraassen him-self takes the crusade against inflationary metaphysics as the pivotal motivation for constructive empiricism (although of course it is a pivotal motivation for his understanding of empiricism in general). In response to Ladyman, Monton and van Fraassen conclude that

> Even if some constructive empiricist were to embrace modal realism – and therefore at least one bit of what van Fraassen counts as inflation-ary metaphysics – she could still argue that constructive empiricism makes better sense of science than realism does. *It is here – regarding how best to make sense of science – that one finds a central, arguably the main motivation, for constructive empiricism.*
>
> (2003: 421 [emphasis added])

The combination of constructive empiricism and modal realism is there-fore both logically consistent and philosophically coherent. However, although there are no *technical* difficulties in combining constructive empiricism with some objectivist account of modality (and given the nature of the beast, certainly no *epistemological* difficulties), the addi-tional ontological commitments associated with modal realism do raise some further problems – problems concerning the similarity ordering of these additional possible worlds, as promised at the beginning of this section. Indeed, Ladyman's (2004a: 764; see also Ladyman and Ross, 2007: 111) fundamental objection to the combination of constructive empiricism and objective counterfactual truth-conditions is not that it is inherently problematic, but that it is in fact a retreat to *struc-tural realism*,[4] which consideration of the similarity ordering of possible worlds makes clear.

The problem can be put succinctly as follows: exactly *which* possible worlds are to be used to delineate what does and does not count as observable? Clearly, the constructive empiricist cannot allow anything that it is logically possible for us to observe to count – to do so would be to collapse the entire distinction between observable and unobservable phenomena, since it is logically possible to observe *any* entity or process, given the right sort of circumstances.[5] The constructive empiricist who endorses objective counterfactual truth-conditions cannot therefore allow unrestricted access to every possible world for the purposes of grounding his distinction between observable and unobservable phenomena – some account is needed to specify the appropriate set of possible worlds. And moreover, once the appropriate set of possible worlds has been determined (presumably on the basis of some suitably restricted notion of nomological accessibility, as determined by our best theories of the behaviour of light and the physiology of the human eye), the constructive empiricist must also provide some account of which of these privileged possible worlds are closer to (i.e., more similar to) the actual world than others if he is to have any way of evaluating the truth-value of his counterfactual conditionals.

What all this amounts to then is that in order for the modal objectivist strategy to work, the constructive empiricist must be committed to more than just the existence of non-actual possibilia, for he must also provide some criteria for privileging one possible world over another when he comes to evaluate his counterfactual conditionals. And unless he wishes to leave this as nothing more than a matter of convention (which would thereby render this foray into objective modality completely redundant – the whole point after all was to avoid any charge of arbitrariness), this means that the constructive empiricist must be committed to some objective similarity ordering of possible worlds. And the problem then is that the additional structure that he must thereby be committed to in order to ground this similarity ordering will be exactly the sort of additional structure that will collapse constructive empiricism into a form of structural realism.

For example, it seems that one important constraint upon the constructive empiricist's similarity ordering will be holding fixed the laws of nature that occur at the actual world – at the very least, when the constructive empiricist is talking about what it is possible for us to observe, he is talking about what it is *physically possible* for us to observe. Thus if the constructive empiricist admits objective modal facts, he must also admit that some of the regularities posited by his accepted scientific theories describe objective features of the actual world that are to be held fixed when judging what counts as the nearest (most similar) possible world. If he did not, then any attempt by the constructive empiricist to endorse objective counterfactual truth-conditions would also face the charge of arbitrariness, since there is no other method available for determining the appropriate possible world for evaluating the counterfactual conditional in question. Countenancing objective truth-conditions for his counterfactual conditionals therefore commits the constructive empiricist to more than the existence of objective modal facts; it also commits him to the objectivity of some of the unobservable structure posited by the models of his accepted scientific theories. And it is in this respect that Ladyman argues that the combination of constructive empiricism and modal realism is better thought of as a version of structural realism.

3.3.3 Deflationism and voluntarism about modality

Endorsing objective counterfactual truth-conditions in order to secure his fundamental distinction between the observable and the unobservable therefore commits the constructive empiricist to just those beliefs that he contends that he need not believe in order to make sense of scientific practice: for if in order to believe that a particular entity or process is observable, the constructive empiricist must not only believe that there is a possible world where that entity or process is observed, but also believe that this possible world possesses an unobservable structure which his accepted scientific theories tell him is also possessed by the actual world, then he is clearly committed to believing his accepted scientific theories to be more than empirically adequate – in direct contradiction to his view regarding the aim of science that such talk of observability was meant to secure. The first horn of Ladyman's dilemma can therefore be seen as a straightforward instance of the general objection raised by Musgrave. Moreover, just as Muller and van Fraassen attempted to rebut Musgrave's original objection by construing their putatively problematic commitments as in fact satisfied by more parsimonious beliefs, so too do Monton and van Fraassen (2003) attempt to grasp the second, deflationary, horn of Ladyman's dilemma. As mentioned briefly above, Monton and van Fraassen defend an essentially meta-linguistic account of counterfactuals, according to which a counterfactual conditional is true iff there is a model of our accepted scientific theories in which both the antecedent and the consequent are true. Their basic strategy is to claim that since the counterfactual conditionals in which we are interested merely describe the logical consequences of our various accepted scientific theories – and since the relevant scientific theory will vary from context to context – there is no sense in which counterfactual conditionals are made true by objective modal facts. However, since these various scientific theories are believed to be empirically adequate, and thus correctly describe certain objective – albeit non-modal – facts about the actual world, using such counterfactual conditionals to determine the distinction between observable and unobservable phenomena is in fact far from arbitrary.

Take for example van Fraassen's account of nomological necessity. In outlining what we might describe as his error theory of laws, van Fraassen (1989) argues that the construction of (the models of) our scientific theories is implicitly constrained by various methodological considerations – symmetry being a particular important and ubiquitous example. These constraints limit the various models that will satisfy a given scientific theory; and when these constraints are sufficiently general, the overall result will be various universal constraints upon our legitimate modelling of the world. According to van Fraassen, it is these universal model-theoretic constraints that encourage philosophers of science to propose robustly metaphysical accounts of the laws of nature – the thought being that if there are these general restrictions upon what counts as a legitimate model of the world, then it is not unreasonable to suppose that this must be due to some fundamental structure of the world itself. But on van Fraassen's view, this is quite misguided: 'laws of nature' are nothing more than the implicit constraints that we have conventionally imposed upon theory construction. To quote van Fraassen at length:

> Laws do appear in this view – but only laws of models, basic principles of theory, fundamental equations. Some principles are indeed deeper or more fundamental than others. Pre-eminent among these are the symmetries of the models, intimately connected with the conservation laws, but ubiquitous in their influence on theory construction. Our diagnosis is *not* that the more fundamental parts of a theory are those which reflect a special and different aspect of reality, such as laws of nature! It is only the content of the theory, the information it contains (and not its structure), which is meant to have the proper or relevant *adequatio ad rem*.
>
> (1989: 188)

For van Fraassen then, laws of nature – and hence claims of nomological necessity and nomological possibility – are merely features of the models of our accepted scientific theories: a claim is nomologically necessary iff it is true in every model of the relevant scientific theory, nomologically possible iff it is true in some model of the relevant scientific theory. And so in general: for van Fraassen, all modal claims are implicitly claims about the logical consequences of our accepted scientific theories.[6] As any good empiricist should, van Fraassen offers 'a robust denial that there are other possible worlds – for possible worlds talk is then only a picturesque way to describe models' (1989: 68); or perhaps more succinctly: 'the locus of possibility is the model, not a reality behind the phenomena' (1980: 202).

What van Fraassen proposes then is in essence the traditional empiricist strategy of reducing the metaphysics of modality to the structure of language, albeit spruced up in various interesting and important ways.[7] The approach is of course model-theoretic rather than syntactic (although in the light of our discussion of Muller and van Fraassen's response to Musgrave, this may well be a difference that makes no difference). For van Fraassen, our modal discourse is also radically context-dependent; he rejects the notion of essences on the grounds that what is to be kept fixed about an individual across different possible worlds will vary depending upon the sorts of questions that are asked about it (van Fraassen, 1981). Moreover, claims about possibility and necessity are also to be understood as tacitly *indexical* (van Fraassen, 1977b), in the sense that one must distinguish between the context of utterance and the circumstances of evaluation of any particular modal utterance.[8] Most of these details however need not concern us in what follows. The basic idea is that since the details of the various contexts from which our modal discourse is to be evaluated will be determined by the models of our various accepted scientific theories, and since as empiricists we should refrain from reifying these models to the status of full-blown possible worlds, it follows that for van Fraassen our modal discourse is indeed made true or false by the models of our accepted scientific theories. Again, as van Fraassen puts it,

> Guided by the scientific theories we accept, we freely use modal locutions in our language. Some are easily explicated: if I say that it is impossible to observe a muon directly, or to melt gold at room temperature, *this is because no counterpart to such events can be found in any model of the science I accept.*
>
> (1980: 201 [emphasis added])

Exactly the same story holds for our counterfactual conditionals. Of course, some counterfactual conditionals do have objective truth-values, purely as a matter of logic. These are all fairly uninteresting examples however, such as a counterfactual conditional of the form $(\varphi \,\square\!\!\rightarrow\, \varphi)$. However, Monton and van Fraassen (2003: 410–411) argue, the truth-value of most other counterfactual conditionals – and certainly the interesting ones about observability – will follow as a matter of logic *relative* to certain background assumptions and scientific theories held fixed by the speaker. To take one of their examples, the counterfactual conditional that if I had looked in the drawer, I would have found the letter is true as a matter of logic *relative* to the tacit premise that the letter is in the drawer (and numerous other premises such that my eyes are working properly, and that the letter doesn't suddenly disappear and so on and so forth). This is the sense in which Monton and van Fraassen argue that most counterfactual conditionals lack objective truth-values and are to be considered as context-dependent (cf. van Fraassen, 1981: 193–195). The context in which a counterfactual conditional is asserted will be given in terms of the various background theories and implicit premises held by the speaker; and when a counterfactual conditional is true, it is true because background theories, together with the antecedent of that counterfactual, logically entail the consequent. Thus what Monton and van Fraassen propose therefore is indeed a meta-linguistic theory of counterfactuals (cf. Chisholm, 1946; Goodman, 1947; Mackie, 1973): counterfactual conditionals do have truth-values – but since these are relative to a context, and since the relevant context will frequently change, Monton and van Fraassen conclude that these modal truth-values are not objectively modal truth-values.

The crucial element in all of this of course concerns the rather slippery notion of a 'context', since the counterfactual conditionals in which the constructive empiricist is interested will have one truth-value relative to one context, and quite another truth-value relative to another. A context within which we allow the physiology of the human eye to be kept fixed, yet which allows my spatio-temporal location to vary, will classify one set of phenomena as being observable; conversely, a context within which my spatio-temporal location is kept fixed, yet which allows the sensitivity of my optic nerves to be vastly increased, will classify an entirely different set of phenomena as being observable. Monton and van Fraassen's deflationary strategy therefore faces the general problem of *cotenability* that plagues all meta-linguistic accounts of counterfactuals (cf. Goodman, 1947), and unfortunately, Monton and van Fraassen have little further to say on the matter: they note that the context within which a counterfactual conditional is uttered will contain 'a good deal of unformed general opinion, but also features specific to the case' (2003: 410), but provide no further details. The problem is again rendered most perspicuously in terms of the similarity ordering of possible worlds (provided of course that in this case, one understands 'possible world' as a picturesque way of describing a model): it is not sufficient for the truth of a counterfactual conditional that there is some model of the relevant scientific theory where the antecedent of the conditional logically entails the consequent; it must also be the model *most similar to* (the model that represents) the actual world. That is to say, it is simply irrelevant to the evaluation of a counterfactual conditional to point out that there is some model of the theory that makes it true, unless that model also manages to keep fixed all of the relevant background information about the case in question.[9]

According to Ladyman (2004a: 762), this is where the real problem with Monton and van Fraassen's account lies. He puts the problem as follows: on what grounds can Monton and van Fraassen maintain that if I were to travel deep into outer space then I would observe the moons of Jupiter? How can they justify privileging this counterfactual conditional over the opposing claim that if I were to travel deep into outer space, then I would not observe the moons of Jupiter on account of them suddenly becoming invisible? Similarly, why suppose that if I were to travel back in time then I would observe the markings upon a particular species of dinosaur, and not that if I were to travel back in time, then said species of dinosaur would disappear or change its appearance? To put the point in terms of models: why is it that when selecting the appropriate model of my accepted scientific theories for evaluating counterfactual conditionals about travelling into outer space, I should choose one that keeps the various facts about the shape, size and constitution of the moons of Jupiter constant, rather than a model that allows these parameters to vary (or when evaluating counterfactual conditionals about travelling back in time, a model that keeps the various facts about the dinosaurs in question constant, rather than allowing them to vary)? Unless Monton and van Fraassen can provide some non-arbitrary account of this – in other words, provide a suitable model-theoretic analogue of the modal realist's similarity ordering of possible worlds – the truth-values of their crucial counterfactuals will remain indeterminate.

The obvious solution of course would be for Monton and van Fraassen to claim that the constructive empiricist can justify privileging one model over another on the grounds that he has reason to believe that the regularities described by our accepted scientific theories describe objective features of unobservable reality – that is, laws of nature. In other words, the reason why travelling deep into outer space would result in my observing the moons of Jupiter – rather than the disappearance of the moons – is that the structural stability of the moons is exactly the sort of substantial fact about the actual world that any sufficiently similar model must represent (and similarly for facts about prehistoric species). By contrast, any model of my accepted scientific theories that predicts the disappearance of the moons – or the sudden visual alteration of the dinosaurs, and so forth, just in virtue of not keeping this substantial fact about the actual world fixed – cannot count as an appropriate model for the evaluation of counterfactual conditionals of this kind.[10]

But such a response is clearly unavailable to van Fraassen who, as we have already seen, is as adamant in his rejection of laws of nature as he is in his rejection of objective modality. According to van Fraassen, 'laws' merely describe part of the implicit structure of the models of our accepted scientific theories: they are facts about how we choose to represent the world, not facts about the world itself. Consequently, the appeal to such devices is quite unable to solve Monton and van Fraassen's problem of cotenability. On this account, to privilege one model of a theory over another, on the grounds that it is nomologically similar to the actual world (i.e., preserves the laws of nature that are said to exist at the actual world), is merely to privilege one model of a theory over another on the grounds that it satisfies more of our *conventions* about how to represent the world. As Ladyman puts it,

> [Monton and van Fraassen] argue that in practice what is observable will follow from certain generalisations about the actual facts. However, my claim is that where X is never actually observed, such generalisations about what actually happens will not be enough to determine anything about what would happen if X was present to someone, unless we take it that the specification by science of some regularities among the actual facts as laws, with the implication that they also hold among non-actualised possibilities, is not merely a matter of pragmatics as van Fraassen argues, but is latching onto objective features of the world.
>
> (2004a: 762)

Ultimately then, Monton and van Fraassen's meta-linguistic account makes the truth of a counterfactual conditional dependent upon certain conventional, pragmatic decisions of the scientific community about how they are to represent the world. In particular then, counterfactual claims about what we would have observed, had the circumstances been different – and thus the constructive empiricist's fundamental distinction between observable and unobservable phenomena – are also dependent upon certain conventional and pragmatic decisions of the scientific community about how they are to represent the world. The problem then of course is that it would seem that the constructive empiricist's distinction between the observable and the unobservable has to be far more substantive than *that* – as I put it earlier, if the distinction is to pass any kind of epistemic muster, it must reflect some kind of objective structure of the world. A distinction based upon little more than our aesthetic sensibilities about modelling, however, really does seem entirely arbitrary.

In order to provide a satisfactory analysis of those counterfactual conditionals vital to his distinction between observable and unobservable phenomena, the constructive empiricist must not only provide the appropriate truth-conditions for his modal discourse, he must also provide some principled criteria for privileging one set of truth-conditions over another. If he takes the deflationary horn of Ladyman's dilemma, and takes our modal discourse to be made true or false by the logical consequences of the models of our accepted scientific theories, it seems that the only method available for privileging one model over another is in terms of the conventions of the scientific community; and this account seems to be far too weak to support the constructive empiricist's position. Yet if he takes the inflationary horn of Ladyman's dilemma, and takes our modal discourse to be made true by what happens at other possible worlds, then it seems that the only method available for privileging one possible world over another is in terms of an objective account of the regularities that hold between the entities and processes represented in his models; and this account commits the constructive empiricist to believing more of the claims of his scientific theories that his view regarding the aim of science allows. The distinction between observable and unobservable phenomena then is either so weak as to render the position quite arbitrary, or so strong as to render it a form of realism.

It is at this point then that we can see just how van Fraassen's considered response to Ladyman hinges upon what appear to be quite arbitrary considerations, just as his considered response to Musgrave ultimately hinged upon what appeared to be quite arbitrary considerations. And similarly, it is at this point that we can see just how crucial an appeal to an epistemologically voluntarist framework becomes. The only option left for Monton and van Fraassen is for them to concede the inherently conventional status of their counterfactual conditionals, yet to challenge the extent to which this renders the constructive empiricist's distinction between observable and unobservable phenomena entirely arbitrary. The whole point of a meta-linguistic account of counterfactuals, Monton and van Fraassen may argue, is after all to dispense with an objective notion of modality in favour of something based in everyday human practice; they might therefore, and perhaps with some justification, complain that in evaluating their deflationary account of counterfactual conditionals against an objectivist standard as Ladyman does – that is, taking the standards of success for a deflationary account of counterfactuals to be to provide what is in effect an objective account of counterfactuals – he somewhat begs the question against them.[11] The question then is whether or not the constructive empiricist's account of counterfactual conditionals needs to satisfy more than just other constructive empiricists. For on the one hand, if we take Ladyman's dilemma as merely a challenge for the constructive empiricist to provide an internally coherent account of his counterfactual conditionals, then perhaps one can simply bite Monton and van Fraassen's deflationary bullet. Yet on the other hand, if we take Ladyman's dilemma as a challenge for the constructive empiricist to provide an internally coherent *and independently plausible* account of his counterfactual conditionals that can also meet his *critics'* standards for a satisfactory resolution of the problem, then Monton and van Fraassen's approach looks woefully inadequate. And while there may indeed be some mileage in the first approach, it is clearly dependent upon construing constructive empiricism as a view about the aim of science, to be situated within a minimalist epistemic framework in which broader questions of justification beyond the basic constraints of logical consistency and probabilistic coherence are redundant.

3.4 Constructive empiricism and mathematical nominalism

3.4.1 Abstract mathematical objects

We have seen then that both Musgrave's objection and Ladyman's dilemma present the constructive empiricist with the rather unfortunate decision between believing so much about the unobservability of unobservable phenomena, or about the truth-conditions of his modal discourse, that he undermines his view regarding the aim of science; or believing so little about these matters that he is unable to establish such a view in the first place. And we have also seen that in both cases, van Fraassen's considered response to these difficulties has been to try and circumvent the deficiencies inherent in the latter, deflationary, strategy by appealing to the minimalist constraints of his idiosyncratic conception of epistemology. In addition to these structural similarities, it has also emerged how Muller and van Fraassen's response to Musgrave in fact depends upon Monton and van Fraassen's response to Ladyman in a rather straightforward manner – for without a satisfactory resolution of the problem of modality in the face of Ladyman's misgivings, Muller and van Fraassen simply lack the resources for making any stipulations concerning the modal scope of the constructive empiricist's various beliefs (even leaving aside any doubts about how such stipulations are to be justified), since without a satisfactory response to Ladyman, the constructive empiricist simply lacks any adequate account of modality about which to make such stipulations. Similarly, Monton and van Fraassen's response to Ladyman is based upon a model-theoretic account of modality which (again, leaving aside any doubts over how exactly the constructive empiricist is supposed to justify the choice of one model over another) therefore depends essentially upon a satisfactory account of the existence of these abstract mathematical objects. And as we shall see in this section, the exact same dilemma over believing too much or believing too little on this matter continues to plague the constructive empiricist in this domain as well.

The final variation on Musgrave's original objection that I wish to consider in this chapter is what we might as well call the problem of mathematics for constructive empiricism. For even without the deflationary account of modality advanced by Monton and van Fraassen in the previous section, there seems to be an important respect in which the constructive empiricist is committed to believing in the existence of certain abstract mathematical objects; while at the same time, given the constructive empiricist's view regarding the aim of science as mere empirical adequacy, belief in such abstract mathematical objects would undermine his position. Constructive empiricism, recall once again, is the view that 'science aims to give us theories that are empirically adequate; and acceptance of a theory involves as belief only that it is empirically adequate' (van Fraassen, 1980: 12). But as Gideon Rosen (1994: 165) argues, for the constructive empiricist to believe that an accepted scientific theory is empirically adequate is for him to believe that the theory in question exists, and that it possesses the complex relational property of empirical adequacy. But if that is indeed the case, then abstract ontological commitment immediately follows since, according to van Fraassen, a scientific theory *is* just an abstract mathematical object.[12]

And unfortunately, just as was the case with the unobservability of unobservable phenomena and with the existence of unactualised possible phenomena, there also seems to be an important respect in which the belief in the existence of abstract mathematical objects is incompatible with constructive empiricism. As Rosen (1994: 164) puts it, constructive empiricism appears to entail a form of *mathematical nominalism*: abstract mathematical objects are unobservable phenomena if anything is, and experience – the source of all of the constructive empiricist's knowledge – is quite unable to tell us whether abstract mathematical objects exist or not. Indeed, van Fraassen (1974) himself has made just this point: in his fable, the two lands of Oz and Id disagree on the existence of sets and other abstract objects – but both live happily ever after *precisely because* the existence of such entities makes no empirical difference. It follows then, according to Rosen, that 'just as [the constructive empiricist] suspends judgement on what his theory says about unobservable physical entities, he should suspend judgement on what they say about the abstract domain' (1994: 164).

Of course, Rosen's presentation of the problem is less than ideal because it implies that since the constructive empiricist does not believe what his accepted scientific theories say regarding unobservable physical phenomena, then he cannot believe what his accepted scientific theories say regarding unobservable abstract phenomena – and as we have seen in our discussion of the problem of modality, not only is such an inference straightforwardly *invalid* (since, again, the epistemological methods involved for knowing about one kind of phenomena will not be the same methods involved for knowing about the other kind of phenomena), it also fails to appreciate that constructive empiricism is an epistemologically neutral view concerning the *aim* of science (and hence compatible with a whole range of different epistemological attitudes regarding a whole range of different phenomena). Nevertheless, Rosen's basic concern is easily reconstrued as a more legitimate objection: if the constructive empiricist believes his accepted scientific theories to be empirically adequate (as he does), and if the constructive empiricist believes his accepted scientific theories to consist of abstract mathematical objects (as he also does), then the constructive empiricist believes in the existence of abstract mathematical objects, for exactly the reasons that Rosen gives. The problem then is not with believing in the existence of abstract mathematical objects *per se*; the problem is that whenever the constructive empiricist believes what his accepted scientific theories say regarding observable physical phenomena, he thereby believes what his accepted scientific theories say regarding the existence of abstract mathematical objects and their relationship to the observable physical phenomena. In which case then, since the constructive empiricist's view about his accepted scientific theories aiming for nothing beyond empirical adequacy actually requires the constructive empiricist to believe that his accepted scientific theories deliver the goods with respect to certain issues that go *beyond* empirical adequacy (i.e., with respect to the existence of the abstract mathematical objects of which these theories consist), his position refutes itself.

Thus again, it seems that the constructive empiricist faces the following familiar dilemma: on the one hand, in order to even state his position, he appears to be committed to believing in what his scientific theories say regarding the existence of abstract mathematical objects; yet on the other hand, in order to remain consistent with the view of science that such beliefs are meant to support, he appears to be committed to *not* believing in what his scientific theories say regarding the existence of abstract mathematical objects – again a dilemma between believing too much and undermining his position or believing too little and being left with a hopelessly impoverished position (in this case, one that cannot even be consistently formulated). And with respect to our preceding discussion, it should also be clear that we have now pursued Musgrave's original objection to its highest level of abstraction: we began by problematising the constructive empiricist's beliefs in the unobservability of unobservable phenomena; generalised this problem to the modal dimensions of unobservability and to the constructive empiricist's modal discourse in general; and in investigating the logical machinery implicit in the constructive empiricist's account of counterfactual conditionals, we have pointed to a tension in the beliefs that the constructive empiricist must hold to even formulate the framework in which the previous challenges took place.

Indeed, the only substantial difference to be found between the dialectic surrounding Rosen's objection and the dialectics surrounding the objections posed by Musgrave and Ladyman is that van Fraassen himself has yet to offer his own explicit response to the problem (either on his own or in collaboration). Monton and van Fraassen (2003: 412) do discuss Rosen's objection, somewhat tangentially, in a footnote. They concede that there is a potential tension between the constructive empiricist's apparent commitment to mathematical nominalism and van Fraassen's explicit use of mathematics in his articulation of constructive empiricism, but argue that the objection only holds on 'the supposition that mathematics is intelligible only if we can view it as a true story about certain kinds of things', which is not a supposition shared by all philosophers of mathematics. This is certainly true, but as a response to Rosen it amounts to little more than the claim that one cannot solve the problems of the philosophy of mathematics *en passant* in the philosophy of science; and as I hope my continued belabouring of Musgrave's problem of internal coherence has shown, along with the associated difficulties with epistemic voluntarism, this is one problem in the philosophy of mathematics that *must* be solved in the philosophy of science if constructive empiricism is to have any future.

Fortunately, the gauntlet has been picked up on van Fraassen's behalf by Otavio Bueno (1999) and his so-called partial structures approach. Bueno's proposal, roughly speaking, is for the constructive empiricist to adopt a *fictionalist* attitude towards the existence of abstract mathematical objects (along the lines originally proposed by Hartry Field), suitably reformulated for empiricist purposes (hence the need for the partial structures, as we shall shortly see). The idea then is that on this basis, the constructive empiricist can argue that his commitment to abstract mathematical objects need not entail the belief in their existence: in fact, the idea is to argue that his commitment to such objects is actually consistent with their non-existence, since the *usefulness* of the various claims he needs to make about abstract mathematical objects can be accounted for in some way other than their *truth*.

Such a response clearly relieves the constructive empiricist from the burden of believing in the existence of abstract mathematical objects since, as with all of the responses considered in this chapter, it argues that such beliefs can in fact be shown to be somehow reducible to a more parsimonious set of beliefs – in effect, beliefs about the consistency of our mathematical discourse, rather than its ontological commitments. Moreover, such a response also seems to be in line with van Fraassen's own thoughts on the matter: in a much earlier piece of work, he notes that he has 'not worked out a nominalist philosophy of mathematics...yet [it is] clear that it would have to be a fictionalist account' (1985: 303). But just as with all of the other responses we have considered in this chapter, it remains to be seen just how successful the combination of constructive empiricism and mathematical fictionalism really is. For despite the addition of Bueno's novel technicalities, the familiar dialectic makes its inevitable appearance, and we see the constructive empiricist's attempt at a deflationary resolution of the problem fails to accommodate everything that it must. Most importantly though, the problem of mathematics is also the end of the road for van Fraassen's systematic deflationism – for as we shall see, the doxastic deficit left by Bueno's approach undermines the very possibility of the equally familiar voluntarist recourse, with catastrophic consequences for every other deflationary response considered in this chapter.

3.4.2 Constructive empiricism and mathematical fictionalism

Bueno's proposed resolution of the problem of mathematics is therefore to 'reformulate in empiricist terms, the fictionalist account put forward by Hartry Field' (1999: S475). As is fairly well-known by now, Field's (1980; 1989) programme within the philosophy of mathematics is an explicit attempt to navigate the two horns of Benacerraf's (1973) dilemma – a problem that has shaped a substantial quantity of the contemporary literature within that domain. The dilemma points to a tension between our intuitive desiderata regarding the *semantics* of our mathematical discourse and our intuitive desiderata regarding the *epistemology* of our mathematical discourse: roughly speaking, a realist semantics for our mathematical discourse (i.e., one that takes our mathematical theories to refer to various abstract mathematical objects) provides the most straightforward account of mathematical truth, but at the expense of postulating referents for our mathematical discourse that it seems impossible for us to know; an epistemologically accessible account of our mathematical theories (i.e., one that reduces such discourse to claims about logical truth, about human mathematical construction and so on and so forth) renders it semantically incomplete. Field's strategy then is to adopt a realist semantics for our mathematical discourse, and thus to provide a satisfactory account of mathematical truth, while attempting to avoid the epistemological problem of how we could ever come to know whether or not our mathematical discourse was true by denying the existence of abstract mathematical objects and thus concluding that in fact all of our mathematical discourse is *false* (or trivially true in the case of universal quantification).

Field's programme therefore provides an ingenious method for resolving Benacerraf's dilemma, but it brings to the fore a new problem concerning the *applicability* of mathematics. To put it bluntly, the mathematical fictionalist needs to provide some account of the successful utilisation of mathematics – in particular within the natural sciences – given that on his view, our mathematical theories are all supposed to be false. In response, Field (1982: 51) argues that the usefulness of our mathematical theories has absolutely nothing to do with their truth. Instead, he argues, it depends upon whether or not such theories are *conservative* – in essence, a strong form of consistency. The idea is that a mathematical theory can be said to be conservative with respect to a particular subject matter iff it is consistent with respect to every internally consistent, non-mathematical theory of that subject matter. More formally, for any theory T that is not itself explicitly mathematical, let

$T^{\#}$ be the result of restricting all of the variables of T so that they are non-mathematical; then a mathematical theory M can be said to be conservative with respect to a particular domain iff for any such T that is a member of that domain, if T is consistent then $(T^{\#} + M)$ is consistent. The basic point is that a conservative theory adds nothing new to the domain in question: $(T^{\#} + M)$ will not derive any non-mathematical conclusions that do not follow from T alone. Nevertheless, a conservative theory may make the facilitation of certain inferences *easier*; and this, according to Field, is where the usefulness of mathematics lies – as an ontologically neutral shorthand in which one can reason about other domains of discourse.

In many ways then, it seems as if a mathematical fictionalist strategy is on the right tracks for resolving the difficulties that Rosen has raised. The constructive empiricist seems to face the dilemma of being committed to believing in the existence of the various abstract mathematical objects posited by his accepted scientific theories in an attempt to establish a position that *denies* that one needs to believe what one's accepted scientific theories say about the existence of abstract mathematical objects. By adopting some suitably formulated version of mathematical fictionalism however, it seems that the constructive empiricist can argue that his belief in what his accepted scientific theories say about abstract mathematical objects need not commit him to believing more than what his explicit view regarding the aim of science licenses, since these beliefs are in fact reducible to far more parsimonious beliefs concerning the *consistency* of what these theories say about abstract mathematical objects, rather than their truth. To help make the preceding structure completely explicit: beliefs about the unobservability of unobservable phenomena are reducible to beliefs about the observability of observable phenomena and (the appropriate modal specification of) the empirical adequacy of the constructive empiricist's theory of observability; beliefs about modal phenomena are reducible to beliefs about the logical consequences of the models of the constructive empiricist's various scientific theories; and beliefs about the existence of the models of the constructive empiricist's scientific theories are reducible to beliefs about the conservativeness of the constructive empiricist's mathematical discourse *about* such models.

Before we can begin to assess Bueno's proposed combination of constructive empiricism and mathematical fictionalism, it is important to recognise that Field's programme actually involves *two* key elements. The first, as we have just discussed, is the emphasis placed upon the notion of conservativeness: since our mathematical discourse is strictly speaking false according to the mathematical fictionalist, Field needs to show that our mathematical theories nevertheless possess some property other than truth that can account for the undisputed range of successful application made of them. The conservativeness of our mathematical theories will count for nought however without the success of the second key element in Field's proposal. As the definition above makes clear, conservativeness can only be employed in conjunction with *non-mathematical* theories: a mathematical theory is conservative with respect to a particular domain iff it is consistent with every internally consistent, non-mathematical theory of that domain. The point of course is that if it is impossible to remove all explicit reference to abstract mathematical objects from all of the theories of a particular domain – that is, if for some theory T of a particular domain, it is *not* possible to restrict all of the variables of T so that they are non-mathematical – then one can hardly defend the applicability of our mathematical discourse to that domain as an ontologically neutral shorthand. What all this means then is that in order to show that our mathematical discourse is indeed conservative with respect to the natural sciences, Field must first of all show that our accepted scientific theories can be successfully *nominalised* – he needs to show that our mathematical discourse is *dispensable* to the natural sciences, and that the scientific theories in which the constructive empiricist is interested can be reformulated without any explicit reference to abstract mathematical objects.

According to Bueno (1999: S479), both of these elements present an initial difficulty for the constructive empiricist. Firstly, an integral part of Field's strategy for the nominalisation of science – and in particular, his strategy for the nominalisation of Newtonian mechanics – involves quantification over space-time points (roughly speaking, Field's nominalisation strategy is to reduce the *quantitative* claims of the theory that appear to make reference to abstract mathematical objects to *comparative* claims that are themselves grounded in various geometric relationships defined over the structure of space-time). This means then that part of Field's programme involves the assumption of substantivalism with respect to space-time, according to which regions of space-time exist in and of themselves regardless of whether or not they are occupied (1980: 34–36; 1989: 171–180). The problem of course is that the constructive empiricist cannot endorse such ontological commitments, since *sui generis* regions of space-time are exactly the sort of unobservable phenomena that the constructive empiricist maintains we need not believe. Or as Bueno puts it, in order for Field to be an *anti-realist* in the philosophy of mathematics, he has to adopt a *realist* attitude in the philosophy of science. And of course, despite Field's arguments for substantivalism, unoccupied space-time regions are precisely the sort of entities an empiricist cannot assume.

(1999: S479)

The first task for Bueno's constructive empiricist reformulation of mathematical fictionalism is therefore to reconstrue Field's nominalisation strategy along more ontologically parsimonious lines.

The second difficulty with the proposed combination of constructive empiricism and mathematical fictionalism concerns the notion of conservativeness. As Field points out, the conservativeness of a theory is clearly a much stronger notion than its mere consistency: one way of seeing this is to note that although one can deduce the consistency of a theory from its truth, one cannot deduce the conservativeness of a theory from its truth (the conservativeness of a theory requiring not only its internal consistency, but also its consistency with respect to a number of other internally consistent theories). Moreover, in addition to being stronger than its consistency, the conservativeness of a theory is also not weaker that its truth: for not only can we not deduce the conservativeness of a theory from its truth, we can also not deduce the truth of a theory from its conservativeness (Field, 1980: 16–19; 1989: 59): a theory can be internally consistent, and consistent with respect to every other internally consistent theory of a particular domain, and still be straightforwardly false – this after all is the whole point of the fictionalist strategy. It follows then, argues Bueno, that 'Field is not countenancing a weaker aim of mathematics [than truth], but only a *different* one' (1999: S476). But again this raises serious difficulties for the constructive empiricist, since, according to Bueno, 'an important feature of the constructive empiricist's axiology is that the aim of science should be *weaker than truth*, otherwise the grounds for adopting an anti-realist proposal are lost' (1999: S478). The same point can be put in terms of the *motivation* for the mathematical fictionalist's countenance of conservativeness: according to the mathematical fictionalist, abstract mathematical objects do not exist; by contrast, however, the constructive empiricist is merely agnostic about their existence. It seems then that as it stands, the mathematical fictionalist's view is just too strong for the constructive empiricist palate – the second task for Bueno's proposed combination of the two positions then is to countenance a more 'diluted' account of the mathematical fictionalist's notion of theoretical conservativeness that can thereby accommodate the constructive empiricist's more liberal epistemic outlook.

It is in response to both of these difficulties that Bueno appeals to the much anticipated device of a *partial structure* – and its attendant notion of *quasi-truth* – as originally formulated in the work of Newton da Costa and Steve French (see da Costa, 1986; da Costa and French, 1989; 1990). In very simple terms, a partial structure provides a rigorous method for accommodating incomplete knowledge of a domain within the standard set-theoretic machinery. More specifically, just as a normal structure can be defined in terms of the various relations of which it consists, so can a partial structure be defined in terms of the various *partial relations* of which it consists; and a partial relation again provides a rigorous method for accommodating incomplete knowledge within the standard set-theoretic machinery. A normal relation consists of two sets of ordered n-tuples: the set of n-tuples for which the relation holds, and the set of n-tuples for which the relation does not hold; a partial relation consists of *three* sets of n-tuples – the set of n-tuples for which the relation holds, the set of n-tuples for which the relation does not hold and the set of n-tuples *for which we do not know* whether the relation holds or not.

The familiar Tarskian account of truth is of course only defined for full structures; thus in order to extend the notion to cover the newly introduced device of a partial structure, the intermediary notion of quasi-truth is also introduced. Again the basic idea is quite straightforward. A partial structure can be extended into a full structure simply by making each of its partial relations determinate: that is, for each n-tuple for which we do not know whether or not a particular relation holds for it, we simply stipulate whether it does or not. A sentence α can then be said to be quasi-true with respect to a partial structure A iff there is a full structure B, where B is a consistent extension of A (i.e., one can extend the partial structure A into the full structure B by a consistent set of stipulations concerning the partial relations of A), and α is true in B in the usual set-theoretic manner.

To return then to the proposed combination of constructive empiricism and mathematical fictionalism. The problem was that the notion of theoretical conservativeness, as defined by Field, was too strong for the constructive empiricist's slightly more permissive epistemological perspective. The crucial point to note then is that the notion of quasi-truth introduced by Bueno is much weaker than truth *simpliciter*: a quasi-true sentence α does not completely describe the domain in question, but merely that aspect of the domain modelled by the relevant partial structure. Thus, since there are numerous different ways in which a partial structure can be extended into a full structure, and since in some of these full structures α may be false, it follows that quasi-truth is weaker than truth: true sentences will be quasi-true (with respect to any structure), but not all quasi-true sentences will be true. Consequently, the notion of quasi-truth can be used to construct a suitably weaker account of theoretical conservativeness. A mathematical theory M is said to be *weakly conservative* if it is quasi-true in a partial structure A with respect to a consistent body N of nominalistic claims (these nominalistic claims impose a basic constraint upon the ways in which a partial structure can be extended into a full structure, by ensuring that no such extension is incompatible with our current empirical evidence). From this it follows that M is weakly conservative iff M is consistent with *some* internally consistent body of claims about the physical world (Bueno, 1999: S482). The notion of weak conservativeness is then much like Field's original notion, with the important exception that weak conservativeness only requires consistency with *some* body of physical claims about the domain in question, whereas Field's notion requires consistency with *all* such bodies of physical claims about the domain.

Bueno thus devotes a lot of effort to ironing out some of the technical difficulties with his proposed combination of constructive empiricism and mathematical fictionalism (see also Bueno 1997 for a detailed account of how to reconstrue constructive empiricism *in general* in terms of the technical notions of partial structure and quasi-truth); it is important however not to lose sight of the forest for looking at the trees. One important issue is whether or not the constructive empiricist is really committed to being agnostic about the existence of abstract mathematical objects as Bueno contends, and which motivated most of our technical detour. Constructive empiricism is a view about the aim of science; and although we have seen how the constructive empiricist appears to be committed to believing what his accepted scientific theories say regarding the existence of abstract mathematical objects in his attempt to formulate a position that *denies* such commitments, there remains something curiously unstable about the fictionalist strategy. In very simple terms, the constructive empiricist is only committed to a particular epistemological attitude regarding the existence of abstract mathematical objects if the existence of abstract mathematical objects is something about which our accepted scientific theories have something to say – if our mathematical discourse is something that can be entirely divorced from our scientific discourse then, since constructive empiricism is merely a view about the aim of science, it is consistent for the constructive empiricist to take any epistemological attitude he likes towards the existence of abstract mathematical objects (compare this with the discussion regarding the possible combination of constructive empiricism and modal realism above). Yet this is *exactly* what the mathematical fictionalist claims: according to Field, our mathematical discourse is completely dispensable to our scientific discourse, our accepted scientific theories can be nominalised so as to remove all commitment to abstract mathematical objects, and the utility of mathematical reasoning is to be relegated to its use as an ontologically neutral device for facilitating inferences. More importantly, the dispensability of our mathematical discourse for the natural sciences is a fundamental component of the fictionalist programme – the usefulness of our mathematical theories is to be accounted for in terms of their conservativeness, and theoretical conservativeness is only well-defined with respect to a successfully nominalised subject matter. In short then, the very criteria for the success of the mathematical fictionalist programme undermines the very motivation for Bueno's combination of mathematical fictionalism and constructive empiricism in the first place – for mathematical fictionalism to succeed, mathematics must be dispensable to science; and if mathematics is dispensable to science, then the existence of abstract mathematical objects poses no significant difficulty for constructive empiricism.

So the first point to note then is that the issue of partial structures and quasi-truth is somewhat of a red herring, since if mathematical fictionalism works *at all*, the constructive empiricist's problem of mathematics is resolved (*modulo* a suitably parsimonious alternative to Field's own choice of nominalisation surrogates) and the project of distilling a more diluted notion of theoretical conservativeness made redundant. The real question then is to what extent mathematical fictionalism (Field's original, full-bodied vintage) really does provide the most appropriate strategy for the constructive empiricist to endorse – as with all of the other strategies examined in this chapter, mathematical fictionalism seeks to resolve a dilemma in the constructive empiricist's doxastic commitments by arguing that a putatively problematic set of beliefs (in this case, beliefs regarding the existence of abstract mathematical objects) are in fact reducible to a more parsimonious set of beliefs (in this case, beliefs regarding the conservativeness of our mathematical discourse). It remains to be seen however whether or not this deflationary approach can secure all that the constructive empiricist needs, and it is to this issue that we now turn.[13]

3.4.3   Primitive consistency and expressive completeness

The central argument that I wish to develop against the current proposal for the constructive empiricist to deflate his abstract mathematical commitments to the more parsimonious belief regarding the conservativeness of our mathematical discourse concerns this crucial notion of theoretical conservativeness (in what follows, Field's original formulation), and in particular the notion of logical consistency upon which it is based. Following Tarski (1936), logical consistency is usually defined in terms of truth in a model, where a set of propositions Γ is logically consistent iff there is at least one model in which all the members of Γ are true. However, a model is an abstract mathematical object – exactly the sort of abstract mathematical object that concerns the constructive empiricist – and therefore according to the mathematical fictionalist does not exist. This then makes the standard definition next to useless: for if there are no models, then there will never be any model in which all the members of Γ are true; and thus no set of propositions could ever be considered as logically consistent. A similar problem faces any attempt to work with a proof-theoretical definition of logical consistency. To say that a set of propositions Γ is proof-theoretically consistent is to assert the non-existence of any formal derivation that begins with a set of propositions of Γ and ends with a contradiction. Yet we clearly cannot produce a concrete token for every possible derivation for every possible set of propositions of Γ. Consequently, we must either allow certain sets of propositions to count as proof-theoretically consistent by default (since as a matter of fact, we have yet to produce a derivation of a contradiction from that set of propositions) or invoke various non-actual derivations – that is, more abstract mathematical objects. It follows then that since the mathematical fictionalist must deny the existence of these abstract mathematical objects as well, a proof-theoretical notion of logical consistency is either radically incomplete or as useless to the mathematical fictionalist as the model-theoretic account with which we began. There is a significant problem then as to what exactly the mathematical fictionalist *means* by a theory being logically consistent – if this can be given in neither model-theoretic or proof-theoretic terms – and hence what it means for a mathematical theory to be conservative.

Field's (1984a) response to this problem is to take the crucial notion of logical consistency to be neither syntactic nor semantic, but rather as a *primitively modal* notion. The basic idea here, following Kreisel (1967), is that we can begin with an intuitive notion of logical implication, taken as a primitive notion in a similar way to that of negation, conjunction and universal quantification. That is to say, the meaning of logical implication is not to be conveyed by way of a definition, but rather by way of specifying the procedural rules involved in its use. So just as we can understand the notion of conjunction simply in terms of the procedure specifying that if we have A and if we have B, then we have (A & B) – rather than, say, in terms of satisfaction in a model or truth-preserving inference – so too can we understand the notion of logical implication in terms of positive rules for recognising successful implication (sufficient for showing, for example, that if there exists a purely formal derivation of A from Γ within the constraints of a typical formalisation of first-order logic, then Γ *implies* A) and negative rules for recognising failed implication (sufficient for showing, for example, that if there is no such derivation, then Γ *does not imply* A).

With this in place, Field (1989: 34–35) then proceeds to use this primitive notion of logical implication to define a one-place sentential operator $\Box_L$ to be read as 'it is logically true that', and defined such that

$$\Box_L(A) =_{df.} (A \vee \neg A) \to A$$

and an inter-definable one-place sentential operator $\Diamond_L$ to be read as 'it is logically possible that', such that

$$\Diamond_L(A) =_{df.} \neg \Box_L \neg(A)$$

Intuitively, since anything which is logically true is therefore true *simpliciter*, and since whenever a proposition is a logical axiom the logical truth of that proposition will also be a logical axiom, Field's sentential operator will obey the following simple rules:

$$\Box_L(A) \supset A$$

$$\Box_L(A) \supset \Box_L(\Box_L(A))$$

as well as a principle of distribution, whereby

$$\Box_L(A \supset B) \supset (\Box_L(A) \supset \Box_L(B))$$

Once we put all of this together, it is clear that $\Box_L$ operates in exactly the same way as does the standard necessity operator $\Box$ within the modal logic S4; and it is in this sense that in taking implication as a primitive notion – to be defined in terms of its operation rather than in terms of proof-theory or model-theory – that Field is led to the idea that implication and related notions such as logical truth are primitively *modal* notions of some kind.

The result of all of this is that instead of making the meta-logic claim that a theory T is logically consistent, with all of its associated abstract mathematical commitments, Field can instead make the more parsimonious object-level claim that the conjunction of the axioms of T is *logically possible*, that is,

$$\diamond_L AX_T$$

where AX_T is the conjunction of axioms of a finitely axiomatisable theory T. The point of course is that logical possibility is here to be understood in terms of (the appropriate formalisation of) our intuitive inferentialist understanding of implication, and thus neutral with respect to the existence of abstract mathematical objects. This then allows the mathematical fictionalist to find an equally parsimonious object-level assertion for the claim that a theory is conservative. Recall that a mathematical theory M can be said to be conservative with respect to a particular domain iff for any non-mathematical theory T that is a member of that domain, if T is consistent then $(T^\# + M)$ is consistent (where $T^\#$ is the result of restricting all of the quantifiers of T so that they are non-mathematical). The primitively modal surrogate for this claim would then be of the form:

$$\diamond_L AX_T \supset \diamond_L ((AX)_T)^\# \ \& \ AX_M)$$

which reads that if the conjunction of axioms of T is logically possible, then so too is the conjunction of axioms of $T^\#$ and M logically possible. Generalising for all such theories, including those that cannot be finitely axiomatised, Field introduces a universal substitutional quantifier Π which takes formulae as its substitution class. Following Gottlieb (1980; see also Field, 1984b), Field then uses this device to represent an infinite conjunction of object-level claims about theories. Thus generalised, a mathematical theory M can be said to be conservative iff

$$\Pi B(\diamond_L B \supset \diamond_L (B^\# \ \& \ AX_M))$$

These technicalities bring us to the crux of the constructive empiricist's response to Rosen: for just as the constructive empiricist's claim that his beliefs regarding the unobservability of unobservable phenomena could be reduced to his beliefs regarding the observability of observable phenomena ultimately rested upon the appropriate modal scope of the empirical adequacy of his theory of observability, and just as the constructive empiricist's claim that his beliefs regarding nomological possibility could be reduced to his beliefs regarding the logical consequences of the models of his scientific theories ultimately rested upon the appropriate similarity ordering of these models, so does the constructive empiricist's claim that his beliefs regarding the existence of abstract mathematical objects can be reduced to his beliefs regarding the conservativeness of his mathematical discourse ultimately rests upon Field's primitively modal account of consistency. And just as we were able to challenge the constructive empiricist as to why it is that the empirical adequacy of his theory of observability enjoys a wider modal scope than the empirical adequacy of any other of his scientific theories, and just as we were able to challenge the constructive empiricist as to how it is that the arbitrary representational conventions of the scientific community can ground our judgements of comparative similarity, so too can we challenge the constructive empiricist as to whether or not a primitively modal account of consistency is sufficiently robust for the uses we would make of it.

The problem can be seen when we note that the mathematical fictionalist's utilisation of a primitively modal operator in discussions of logical consistency invites an immediate comparison with similar strategies in the treatment of modality in general. For in much the same way that Field rejects the need for a model-theoretic semantics for our understanding of our mathematical discourse, so too have various philosophers (see, e.g., Fine, 1977) rejected the need for a possible worlds semantics for our understanding of our modal discourse: they argue that our modal notions are to be given in terms of primitive sentential operators in much the same way that Field argues that our logical notions are to be given in terms of primitive sentential operators; and that our talk of and commitment to other possible worlds is to be rejected in much the same way that Field rejects our talk of and commitment to abstract mathematical objects.

However, there is a serious problem of expressive adequacy for modal languages that do not quantify over possible worlds: the argument is that there are some intuitively true modal claims that cannot be expressed without some kind of possible worlds semantics – and therefore, since we appear to be committed to the truth of such claims, and since these claims can only be expressed by quantification over possible worlds, we must therefore be committed to the existence of such worlds. As we shall see, this argument readily generalises to Field's primitively modal notion of consistency; the upshot then is that since Field's mathematical fictionalism is radically incomplete with respect to our basic meta-logical notions, any attempt by the constructive empiricist to absolve himself of abstract mathematical commitment by endorsing such a strategy will similarly leave him with a radically impoverished position – and in just the same way as the other deflationary strategies canvassed in this chapter.

Consider, for example, the modal claim:

1. There could have been more things than there actually are.

This is clearly an example of an intuitively true modal assertion, and one that we would expect any satisfactory account of our modal discourse to capture. And indeed, it is easily expressed in the kind of two-sorted extensional language (i.e., a language that quantifies over both possible objects and possible worlds) familiar to any advocate of possible worlds semantics, in the form

$$\exists w[\forall x(Exw^* \rightarrow Exw) \,\&\, \exists y(Eyw \,\&\, \neg Eyw^*)]$$

where E is a two-place predicate for 'exists at', which takes possible object constants/variables and possible worlds constant/variables in its first and second places, respectively, and where w^* designates the actual world.

However, as Alan Hazen (1976) originally showed, modal claims such as (1) cannot be expressed in a language that uses only the primitive modal sentential operators $\Box$ and $\Diamond$.[14] For on the one hand, in order to assert the existence of an individual at a possible world other than the actual world, it is necessary to use a quantifier inside the scope of a modal operator (thereby allowing us to assert that it is possible that there exists such-and-such an individual). Yet on the other hand, in order to assert that such an individual is distinct from any individual existing at the actual world (i.e., to make a claim of comparative possibility), it is necessary to use a second quantifier which as a matter of *syntax* would also be inside the scope of the first modal operator (in order to be compared with the first quantifier, which is clearly governed by the modal operator), yet which as a matter of *semantics* cannot be governed by that operator (since the domain of this second quantifier is the actual world, and not the domain introduced by the modal operator). For example, we might try to formulate (1) as

$$\Diamond[\forall x(AEx \rightarrow Ex) \& \exists y(\neg AEy)]$$

where A is the *actuality operator*, such that $A\varphi$ is true in a model iff φ is true at the actual world in that model. In this case, we get the correct comparison between the two quantifiers, since it tells us that there is an individual within the domain of quantification that does not exist at the actual world. But unfortunately, since the universal quantifier is within the scope of the possibility operator, its domain is restricted to the possible world introduced by that possibility operator. Consequently, this formalisation no longer tells us that there is a possible world that contains everything that the actual world contains, plus something more (a plausible reading of the claim that there could have been more things than there actually are); rather, it tells us that there is a possible world at which, for everything that exists at the actual world – and which already exists at this possible world – it exists at this possible world, and something exists at this possible world that does not exist at the actual world (the effect of the modal operator is therefore to restrict the domain of any quantifier that falls under its scope). And this is clearly not the content of the claim that there could have been more things than there actually are, since such a formalisation can be satisfied by any possible world that contains a tiny subset of what exists at the actual world, plus one non-actual individual.

Conversely, we could try to avoid this problem by unrestricting the domain of the universal quantifier with the addition of another modal operator:

$$\Diamond\{[\Box\forall x(AEx \rightarrow Ex)] \& \exists y(\neg AEy)\}$$

This now reads as 'necessarily, for any x . . . ', thus referring to all possible individuals, not just those introduced by the initial possibility operator. Unfortunately, in so doing we are no longer able to assert the appropriate comparison between the two quantifiers: since the entirety of the first conjunct is governed by the necessity operator (which in effect trumps the first operator), while the second conjunct remains governed by the original possibility operator, the formalisation as a whole merely tells us that it is necessarily the case that everything that exists at the actual world exists, and that there is a possible world that contains at least one non-actual individual. But such a claim is even easier to satisfy than the first, since it merely requires there to be at least one possible individual – by introducing the second modal operator, it is no longer the case that the two quantifiers range over the same domain.

Thus (1) is an example of an intuitively true modal claim that it appears cannot be expressed without recourse to a possible worlds semantics – the basic point being that without the additional semantic resources of a quantifier notation (which allows a finer-grained specification of scope through its specification of a domain), we lack the capacity to make claims of comparative possibility. It seems then that any satisfactory modal language must be committed to the existence of possible worlds in some sense. Most importantly for our present purposes, an entirely analogous argument can be run against Field's mathematical fictionalism, with obvious consequences for the constructive empiricist who endorses such a programme. For it seems that there are intuitively true claims that we can make regarding *logical possibility*, yet which cannot be expressed in a language that does not quantify over *models*, for example:

2. It is *logically consistent* that there be more things than there actually are.

Clearly, (2) is an example of an intuitively true claim regarding logical consistency that we would wish any satisfactory account of our meta-logical discourse to capture: after all, it is logically consistent that there should be all of the things that there actually are, and there is nothing inherently inconsistent about adding one more. And it should be equally clear that such a claim can be easily expressed in two-sorted language that allows quantification over model-theoretic objects, in exactly the same way that (1) can be expressed in a two-sorted language that allows quantification over possible worlds (in this case, we would be interested in formalising the idea that for any given mathematical model, there is a model of greater cardinality). Finally, it should also be clear that for exactly the same reasons that (1) cannot be expressed within a primitive modal language, neither can (2) be expressed in terms of Field's primitive modal operators. The issue of expressive adequacy therefore poses a serious challenge for mathematical fictionalism, and its proposed combination with constructive empiricism. For if there are meta-logical claims like (2) that we would demand any satisfactory analysis of our meta-logical discourse to capture, and if such claims are inexpressible without quantification over model-theoretic objects, then the mathematical fictionalist is wrong to claim that he can avoid commitment to the existence of abstract mathematical objects; and more importantly, the constructive empiricist is thereby wrong to think that such a deflationary approach to the problem of mathematics will secure all of the beliefs that it must.[15]

3.4.4 The vices of voluntarism

The problem with which we began this discussion of mathematical fictionalism concerned what appeared to be the constructive empiricist's commitment to what his accepted scientific theories said regarding abstract mathematical objects, in direct opposition to his explicit view concerning the aim of science. Although van Fraassen himself has not addressed this particular variant on Musgrave's original objection, we have now examined at length what would appear to be the obvious response – Bueno's proposal that the constructive empiricist adopt some version of mathematical fictionalism with respect to these putatively problematic beliefs. For not only does van Fraassen himself hint at various junctures his own sympathy for mathematical fictionalism, it should also be clear just how closely such a strategy mirrors the constructive empiricist's responses to both Musgrave and Ladyman – for in essence, the mathematical fictionalist's claim is that our beliefs regarding the existence of abstract mathematical objects are in fact reducible to far more parsimonious beliefs regarding the conservativeness of our mathematical discourse. And what I hope the preceding discussion has also made clear is just how the mathematical fictionalist's response must inevitably suffer the same fate as the constructive empiricist's responses to Musgrave and Ladyman.

The crux of the issue turned upon the notion of logical consistency: for just as the constructive empiricist found that his beliefs regarding the observability of observable phenomena and the empirical adequacy of his theory of observability failed to capture the full modal scope of his beliefs regarding unobservable phenomena, and just as the constructive empiricist found that his beliefs regarding the logical consequences of the models of his accepted scientific theories failed to capture the appropriate notion of similarity fundamental to his beliefs regarding counterfactual conditionals, so too has the constructive empiricist found that his beliefs regarding the conservativeness of his mathematical discourse – committed as they are to a primitive inferentialist account of implication – fail to capture his beliefs regarding the notion of logical consistency. We have seen that mathematical fictionalism does not have the technical resources to provide an adequate account of this basic logical concept, for there are some claims about logical consistency that the constructive empiricist would wish to hold as true but which cannot even be expressed within the mathematical fictionalist's primitive modal language; in fact, it appears that such claims can *only* be expressed within a language that explicitly quantifies over the very abstract mathematical objects that the strategy was attempting to avoid.

Thus again, the deflationary response to the general dilemma leaves the constructive empiricist with a hopelessly impoverished position – for let us not underestimate the ubiquity of the logical notions that lie beyond the mathematical fictionalist's grasp. For all of its similarities with the problem of unobservability and the problem of modality, there is however a significant difference with the problem of mathematics – for unlike the first two problems considered in this chapter, there can be no voluntarist solution for the constructive empiricist's deflationary approach to his abstract mathematical commitments. When dealing with Musgrave's objection regarding the constructive empiricist's beliefs regarding the unobservability of unobservable phenomena, Muller and van Fraassen were able to make up the doxastic deficit between the belief that there are no actual observable electrons and the belief that electrons are unobservable, by simply stipulating the appropriate modal scope of the constructive empiricist's beliefs and justifying such a manoeuvre on the basis that they were working within an epistemological framework that demands nothing more of a position than its logical consistency and probabilistic coherence. Similarly, when dealing with Ladyman's dilemma over the constructive empiricist's account of counterfactual truth-conditions, Monton and van Fraassen were able to justify their claim that a purely conventional approach to the similarity ordering of the models of their accepted scientific theories was sufficient for the constructive empiricist's needs, since again such a response had the virtues of logical consistency and probabilistic coherence and they denied that there need be anything more to add. But with respect to Rosen's concern over the constructive empiricist's commitments to what his accepted scientific theories say regarding the existence of abstract mathematical objects, the constructive empiricist could hardly argue that the restricted notion of logical consistency forced upon him in an effort to provide a deflationary account of theoretical conservativeness could be augmented in any way that meets the epistemic voluntarist's minimal requirements of logical consistency and probabilistic coherence, since it is precisely those logical notions that the constructive empiricist is attempting to secure! Epistemic voluntarism cannot solve the problem of mathematics for the constructive empiricist, and the dilemma remains: either believe what his accepted scientific theories tell him regarding the existence of abstract mathematical objects and thus undermine his explicit view regarding the aim of science or endorse a deflationary approach to his mathematical theories and be left with an incomplete and impoverished account of one of our most basic logical concepts.

But if the constructive empiricist cannot solve the problem of mathematics, then he cannot solve the problem of modality either – for we have seen how Monton and van Fraassen's deflationary account of counterfactual truth-conditions explicitly appeals to a model-theoretic account of modality, and is thereby hostage to the constructive empiricist's failure to provide a satisfactory account of such objects. And if the constructive empiricist cannot solve the problem of modality, then he cannot solve Musgrave's original objection after all – for in making their stipulations regarding the modal scope of the constructive empiricist's beliefs about the empirical adequacy of his theory of observability, Muller and van Fraassen explicitly appeal to the constructive empiricist's account of modality in general and are thereby hostage to Monton and van Fraassen's failure in this respect. Epistemic voluntarism has therefore been exposed for the illusion that it is: it has been responsible for nothing more than the artificial preservation of an implausible methodology in the constructive empiricist's response to the various permutations of Musgrave's objection, which when pursued to a sufficient level of abstraction must finally give up the ghost. Since van Fraassen's minimalist epistemology is therefore perfectly impotent with respect to the issue of internal coherence, it is time to reconsider the articulation and defence of constructive empiricism altogether.

3.5 Summary

In Chapter 2, we saw many reasons to be suspicious of van Fraassen's epistemic voluntarism. The various arguments put forward to motivate such a position were found seriously wanting: considerations of diachronic probabilistic coherence seemed to *presuppose* a voluntarist epistemology, rather than providing compelling reasons for its adoption; the sceptical case against a traditional, rules-based epistemology transpired to be too sceptical to license any epistemological alternatives; and it remained far from clear that the proper understanding of empiricism, and its 'recurrent rebellion' against metaphysical speculation, required the perspective of an epistemic stance. Moreover, there remained a number of respects in which epistemic voluntarism failed to tally with our intuitive conception of rationality, in particular in that it appeared to license a form of epistemological relativism. Perhaps most importantly for our concerns however is a continuing unease with the role epistemic voluntarism plays in van Fraassen's articulation and defence of constructive empiricism, most prominently with respect to the epistemic relevance of his central distinction between observable and unobservable phenomena – a role that seems to license the cheerful evasion of any substantial criticism of the position (although so far, such substantial criticisms have not been forthcoming), on the grounds that the only criteria that any philosophical defence needs to meet are the minimal ones of logical consistency and probabilistic coherence.

None of these considerations of course are conclusive, and in this chapter I have therefore pursued a different and potentially more damaging line of argument – that regardless of our general misgivings about van Fraassen's minimalist epistemology, it is at least clear that the *constructive empiricist* cannot be content with epistemic voluntarism. I began with Musgrave's objection to the effect that constructive empiricism is internally incoherent, on the grounds that there are certain beliefs that the constructive empiricist needs to hold in order to draw his central distinction between observable and unobservable phenomena that are in fact rendered illegitimate by the very view of the aim of science that such a distinction was meant to establish. I then considered Muller and van Fraassen's response to this objection, and showed how its plausibility rested solely upon the endorsement of the minimal criteria of success associated with epistemic voluntarism.

Given the otherwise arbitrary nature of Muller and van Fraassen's response, and given that the basic function of epistemic voluntarism within that response was merely to finesse the charge of arbitrariness on the grounds that it presupposes an old-fashioned and discredited epistemological framework, the constructive empiricist seemed to be left in a rather unattractive position regarding both the internal coherence of his philosophy of science and the overall plausibility of his epistemic framework; but again, such considerations could not be taken as conclusive. Muller and van Fraassen's response, however, concerned the modal scope of the constructive empiricist's beliefs regarding the empirical adequacy of his theory of observability, and so the argument proceeded to the constructive empiricist's account of modality in general. And here we saw the same dilemma – and the same subsequent dialectic – that underlay Musgrave's objection repeat itself in a series of objections raised by Ladyman over the constructive empiricist's counterfactual truth-conditions; and again, we saw how the plausibility of

Monton and van Fraassen's deflationary response rested solely upon the wholesale endorsement of the minimal epistemological standards associated with van Fraassen's epistemic voluntarism.

The argument was therefore pursued to one further level of abstraction – since Monton and van Fraassen's deflationary account of modality was based upon the logical consequences of the models of the constructive empiricist's accepted scientific theories, we turned to consider the constructive empiricist's account of abstract mathematical objects. Again, Musgrave's objection resurfaced in a challenge posed by Rosen; again we followed the constructive empiricist down the deflationary horn of the dilemma (this time in his endorsement of mathematical fictionalism); and again we found the constructive empiricist's attempt to reduce his putatively problematic beliefs to a more parsimonious set of beliefs fail to capture all that it needed to. It is at this point however that the voluntarist defence finally collapses. At the level of the constructive empiricist's abstract mathematical commitments, there simply is no room for the characteristic tactic of attempting to make good upon one's doxastic deficit by way of an arbitrary epistemological stipulation to be justified on no other grounds than its logical consistency and probabilistic coherence, since the very beliefs that the constructive empiricist needs to secure are exactly those concerned with the notion of logical consistency itself. The voluntarist defence at this point therefore runs itself into a circle, since one can hardly *justify* one's response on the grounds of its logical consistency when what one's response is *trying to provide* is the very notion of logical consistency to which one is trying to appeal.

The voluntarist defence therefore fails at the level of abstract mathematical commitments. But if the constructive empiricist cannot provide a satisfactory account of the models of his accepted scientific theories, then he can hardly defend a model-theoretic account of modality and a meta-linguistic account of counterfactual truth-conditions; so the voluntarist defence fails at the level of modality too. And if the constructive empiricist cannot provide a satisfactory account of modality, then the problem will trickle all the way back down to Musgrave's original objection too, since the constructive empiricist can hardly make a stipulation about the modal scope of his beliefs regarding the empirical adequacy of his theory of observability if he does not have a satisfactory account of modality to underpin it all.

Musgrave's objection, in all of its forms, remains unsolved; and epistemic voluntarism – in addition to being both poorly motivated and an implausible account of rationality – is unable to make any positive contribution to the articulation and defence of constructive empiricism. It is therefore time to dispense with epistemic voluntarism altogether (or at the very least, to bracket its role in the articulation and defence of constructive empiricism) and to re-consider the constructive empiricist's view regarding the aim of science.

3.6   Appendix – An expanded modal language

One possible solution to the problem of expressive adequacy is to expand the primitive modal language with which the mathematical fictionalist has to work with. In the case of modality, by far the most extensive work in this field has been undertaken by Graeme Forbes (1989). Following Peacocke (1978), the basic idea is to introduce a denumerable number of subscripted operators of the form $\diamond_n$, $\square_n$ and A_n. These are best understood in terms of their model-theoretic clauses. We first extend the model for the language to include an ordered sequence of possible worlds w, where $w[v/i]$ is the result of substituting v for the ith world in the sequence w, and w_n is the nth member of w. Then, when evaluating a formula governed by $\square_n$ at some world w, the subscripted modal operator is understood as sending us to some other world w' to evaluate the sub-formula governed by the operator, and as storing w' in the nth place of w. If the sub-formula is governed by A_n, the sub-formula is evaluated from the world w'. The overall effect of these new operators is to introduce an extra dimension of the scope to our primitive modal language – since each operator now comes with its own specification of the possible worlds appropriate to its evaluation, we no longer have the problem of embedded quantifiers having their domains restricted by whatever comes before them in a formalisation.

For example, such an augmented language gives Forbes the resources he needs to express (1), which he formalises as

$$\diamond_1\{[\square\forall x(AEx \rightarrow A_1Ex)] \,\&\, \exists y(\neg AEy)\}$$

which can be interpreted as follows. As before, we have unrestricted the universal quantifier by prefixing it with a necessity operator. However, whereas before this necessity operator dominated the rest of the sub-formula, the new subscripted actuality operator tells us that the second existence predicate is to be evaluated with respect to the possible world introduced by the initial possibility operator (which has the same index). The first conjunct therefore no longer asserts the rather trivial claim that every possible world necessarily contains what it contains; rather, the first conjunct now tells that there is a world that contains every individual that the actual world contains. Moreover, since the second conjunct is still straightforwardly under the scope of the initial possibility operator, it still tells us that at this possible world that contains everything that the actual world contains, there is in addition an individual that the actual world does not contain; in sum, that there could well have been more things than there actually are. Thus this expanded modal language, by explicitly introducing a device for specifying the individual scope of individual modal operators, can express those claims that escaped a more basic language.

The obvious proposal then would be for the mathematical fictionalist to avoid his similar problem of expressive adequacy by similarly expanding his primitive modal language to include subscripted operators of the form $\diamond_{Ln}$, $\square_{Ln}$ and A_{Ln}. But before the constructive empiricist draws too much comfort from this development, there are a number of concerns that can be raised regarding how exactly this expanded modal language achieves its results. As Melia (1992; 2003: 89–97) has argued, there may in fact be a sense in which Forbes' strategy actually invokes the very possible worlds that it seeks to do without. He argues, for example, that in a case of genuine ontological reduction by paraphrase, the purported paraphrase (in this case, Forbes' primitive modal language and the meta-logical variant under consideration on behalf of the mathematical fictionalist) displays a significant difference in syntax to the original formulation. To take a simple example, the sentence

there are (exactly) two dogs

can be paraphrased so as to eliminate any reference to numbers:

$$\exists x \exists y [Fx \,\&\, Fy \,\&\, x \neq y \,\&\, \forall w(Fw \to w = x \lor w = y)]$$

where F is the property of being a dog. Such a paraphrase clearly displays different syntactic characteristics from the original sentence. Yet as Melia notes, Forbes' expanded primitive modal language actually displays remarkable structural similarities to the kind of two-sorted extensional language that it supposedly opposes. Consider again the quantified formulation and the subscripted primitive modal operator formulation: the only real noticeable difference between the two is that the second formulation uses an actuality operator, AEx, instead of a possible worlds variable, Exw^*; and a subscripted possibility operator, $\diamond_1$, instead of a possible worlds quantifier, $\exists w$. This is made particularly apparent by placing the two alleged alternatives next to one another:

$$\exists w[\forall x(Ex\mathbf{w}^* \to Exw) \& \exists y(Eyw \& \neg Ey\mathbf{w}^*)]$$

$$\Diamond_1 \{[\Box\forall x(AEx \to A_1Ex)] \& \exists y(\neg AEy)\}$$

The similarities are made even more explicit when we come to construct a translation schema between the two. If we were to write out each of Forbes' purported reductions in full, including the actuality operator and subscripting all of the operators even when such completeness would be redundant, we see that a sentence of the form

$$\ldots \Diamond_n (\ldots A_n Ra_1 \ldots a_m \ldots) \ldots$$

can be translated in the form

$$\ldots w_n(\ldots w_n Ra_1 \ldots a_m \ldots) \ldots$$

while a sentence of the form

$$\ldots ARa_1 \ldots a_m \ldots$$

can be translated in the form

$$\ldots w^* Ra_1 \ldots a_m \ldots$$

The worry is now apparent, for it seems as if $\Diamond_n$ *simply is* $\exists w_n$, that A *simply is* w^* and that A_n *simply is* w_n. At the very least, one might argue, the burden of proof is upon Forbes to show that his primitive modal language is not simply a notational variant on its extensional rival – in other words, that he isn't secretly quantifying over possible worlds after all.

Another way to put the problem is to note how exactly it is that Forbes' subscripted operators manage to solve the problem of expressive adequacy. The advantage of using a quantifier notation over that of just using ordinary sentential operators is that a quantifier can specify its own domain, whereas sentential operators will always influence those that follow; and in very crude terms, the way in which a quantifier manages to do this is in terms of its explicit ontological commitments – each quantifier specifies a particular domain of objects, and then makes a claim about those objects. So it is the deliberate imposition of *semantics* onto the two-sorted extensional language that provides the additional dimension of scope unavailable to the formalisation considered purely syntactically. And on the face of it, Forbes' account exactly mirrors this proposal – his subscripted operators add an extra dimension of scope by specifying which sub-formulae are to be paired with which sub-formulae in terms other than their syntactic ordering. But while we have a perfectly good understanding of how this mechanism works for a quantified language – ontological commitment provides the additional content needed to make these distinctions – no such mechanism is offered in support of Forbes' expanded modal language. Thus without such an explanatory story, the suspicion must be that the way in which a subscripted sentential operator designates scope will be the same way in which a quantifier designates scope – and thus that these subscripted operators are simply quantifiers in disguise.

There are good reasons therefore to suggest that Forbes' translation does not constitute a genuine ontological reduction by paraphrase, and that therefore the Hazen/Melia conjecture is correct that an expressively adequate modal language (and by extension, an expressively adequate logical language) must be ontologically committed to possible worlds (and by extension, abstract mathematical objects). The apparent equivalence of the two translations alone however is not sufficient to refute Forbes' contention, for it still remains to be shown *which* formulation has priority. It could well be the case that it is the two-sorted extensional formulation that is an attempted ontological *inflation* by paraphrase of Forbes' more parsimonious formulation – or in other words, just because possible worlds quantifiers and subscripted modal operators are in fact the same linguistic device, it may still be the case that ontological commitment to possible worlds is an unnecessarily extravagant way to understand a perfectly sensible semantic operation. But as the previous considerations concerning how exactly Forbes' proposal works make clear, this seems difficult to maintain. Consider again the way in which we come to understand the additional operators that Forbes employs. These are initially introduced by way of a possible worlds model-theory, one which Forbes claims is simply an ontologically neutral way for assigning truth-values to modal sentences. This therefore implies that we have some kind of independent grasp of these operators prior to the introduction of the model-theory. But this seems incredible. Consider a sentence of the form '$\diamond_1\varphi$'. This is to be understood as telling us to go to some possible world w, store that world in the first position of an ordered sequence of possible worlds w and then see if φ is true at that first possible world w. As Melia argues (1992: 44–47), there seems to be no way of understanding this without the model-theory. Forbes' formulation therefore cannot be understood independently of the extensional formulation; thus Forbes' formulation cannot have priority over the extensional formulation – in fact, it seems to be epistemically parasitic upon a previously understood possible worlds semantics. It seems to be the case then that we should understand Forbes' formulation as the attempted paraphrase, and given the reasons we have seen to suppose that such a paraphrase fails, it would appear that Forbes' allegedly more parsimonious modal language actually invokes the possible worlds it attempts to do without.

It follows quite straightforwardly then that any attempt by the constructive empiricist to resolve Rosen's objection through the adoption of mathematical fictionalism will be vulnerable to exactly the same difficulties. Any subscripted primitive modal operator used to capture various intuitively true meta-logical claims such as (2) would also display the same syntactic structure as a model-theoretic quantifier. Moreover, the introduction clauses for these operators would only be comprehensible if presented within a rich model-theory; and it would be equally mysterious just how exactly these operators can indicate their appropriate scope if they are not employing the same (ontologically committing) mechanism as a model-theoretic quantifier. Any such purported extension of the mathematical fictionalist's primitive modal language would therefore be parasitic upon the notion of quantifying over abstract mathematical objects, and would therefore be entirely unacceptable to the mathematical fictionalist. Indeed, in many ways the problem here is actually worse for the mathematical fictionalist. For what the Hazen/Melia argument shows in the case of modality is that an expressively adequate modal language must be committed to a possible worlds *model-theory*: it would therefore be open to the primitive modalist to continue to deny the existence of other possible worlds, provided he was willing to endorse the existence of abstract mathematical objects (it is an argument for modal realism, but not necessarily the sort of *genuine* modal realism advocated by David Lewis). Such a concession is clearly unavailable to the mathematical fictionalist however: for in the case of logical consistency, it is the existence of abstract mathematical objects themselves that is at issue.[16]

A final response that I wish to consider for the mathematical fictionalist (and thus for any constructive empiricist considering endorsing mathematical fictionalism) is whether or not there is some other way of using a model-theoretic semantics without endorsing any of the ontological commitments associated with it. Indeed, Field has proposed something along these lines in an attempt to account for the applicability of mathematics in meta-logical reasoning; and so it seems worth pursuing the possibility of extending this strategy to accommodate our more general problem of expressive adequacy.[17]

Field argues that the purpose of a model-theory is simply to elaborate upon the notion of logical possibility, which it does via the following two schemata. The *model-theoretic possibility* (MTP) schema states that

If there is a model for A, then $\Diamond_L(A)$

while the *model existence* (ME) schema states that

If there is no model for A, then $\neg\Diamond_L(A)$

(contraposing the latter, these two schemata of course amount to the claim that there is a model for A iff $\Diamond_L(A)$). As they stand of course, these two schemata are useless for the mathematical fictionalist: since on his view that are no such things as models, MTP will always be trivially true since its antecedent is always false; and ME will be useless since its antecedent will be true regardless of whether its consequent is true or false.

Field argues however that the mathematical fictionalist can avoid these difficulties by proposing nominalistically acceptable deflationist surrogates for the schemata in question. The idea is that a mathematical fictionalist can use a model-theory provided it can be shown to function merely as a useful shorthand, in much the same way that the mathematical fictionalist proposes to account for the applicability of mathematical reasoning within the natural sciences. In effect, the proposal is for a kind of second-order mathematical fictionalism: one attempts to account for the applicability of our mathematical discourse within the natural sciences by showing that it is nothing more than a conservative extension of the natural sciences; unfortunately, in order to make sense of this notion of theoretical conservativeness, the mathematical fictionalist appears to be committed to exactly the sort of abstract mathematical commitments that he was trying to avoid insofar as any satisfactory account of consistency seems to involve ontological commitment to models; the idea then is to try and make sense of the applicability of our model-theoretic discourse – within the context of trying to make sense of the applicability of our initial mathematical discourse – in much the same way, thus without incurring its ontological commitments.

According to Field, all that needs to be claimed in order to reap the benefits of a model-theory is the conditional claim that *if* standard mathematics entails a model for A, then A is logically consistent. Since a conditional of this form makes no claim about whether or not standard mathematics is true, or indeed whether or not such models do actually exist, accepting such claims carries no ontological commitment. Consequently, the *deflationist* model-theoretic possibility (MTP#) schema states that

If $\Box_L$(mathematics is true $\supset$ there is a model for A), then $\Diamond_L(A)$

and similarly, the *deflationist* model existence (ME#) schema states that

If $\Box_L$(mathematics is true $\supset$ there is no model for A), then $\neg\Diamond_L(A)$

The problem of course is that regardless of whether or not such a strategy can account for the applicability of model-theoretic reasoning in the mathematical fictionalist's meta-logic, such a strategy as it stands is not going to help solve the problem of expressive adequacy. The kind of claim that needs to be accounted for is not the right kind of expression for either MTP# or ME#: it is not that the mathematical fictionalist needs to posit the existence of a model to discover whether or not such a claim is consistent, but rather that the mathematical fictionalist needs to posit the existence of models in order to even *express* the sort of claim in question. The mathematical fictionalist may be able to utilise a model-theory to show that A is logically possible without being committed to abstract mathematical objects, but he would still have the problem of not being able to formulate the content of A without the use of subscripted modal operators, or model-theoretic quantification.

What the mathematical fictionalist would have to do then is to find some way of utilising the sort of deflationary schema proposed by Field in order *to give the meaning of (the mathematical fictionalist's version of) Forbes' subscripted operators* in such a way that does not depend upon the existence of abstract mathematical objects. A sketch of such a schema, which attempts to give clauses for the subscripted operators in terms of the less contentious unsubscripted operators, would presumably be of the form:

If $\Box_L$(mathematics is true $\supset$ there is a model m, a model m' such that $\Diamond_L(A)$, and an ordered sequence of models *m*, such that m' is in the *n*th place of *m*), then $\Diamond_{Ln}(A)$

the idea being of course that when Forbes gives his model-theoretic clauses for the introduction of his subscripted modal operators, all that really needs to be the case is some conditional claim regarding whether or not the hypothetical truth of this model-theory would guarantee the appropriate semantic structure.

It is plausible then that a series of schemata along these lines would succeed in at least giving clauses for the introduction of subscripted modal operators without invoking the very model-theory that it seeks to replace. But this is only one half of the problem: crucially, we still need to know what these subscripted operators *mean*. Indeed, the only reason that the original schemata that Field introduced for the unsubscripted modal operators could be considered as anywhere near acceptable was that there was supposed to be an independent understanding of these operators, based upon a primitive inferentialist notion of implication. The schemata were justified only insofar as they were held to follow from the meaning of the unsubscripted operators, not the other way around. But such a claim cannot be made in the case of the subscripted operators: we have no independent understanding of what they mean and thus the mathematical fictionalist cannot even justify his acceptance of such schemata, let alone invoke them in order to clarify his modal terminology.

In fact, whichever way the mathematical fictionalist looks at it, he is onto a loser. He cannot construct deflationary schemata to give him the meaning of his primitive modal operators since such schemata are only acceptable if the operators they introduce are already understood. But even if he *could* appeal to such schemata, he would undermine his position. The problem is that, for the mathematical fictionalist, model-theoretic reasoning can be accounted for only so far as it serves as a useful shorthand. It is for finding out about logical possibility, not for giving its content. But by the very fact that the model-theory is being used to explain these primitive modal operators, the mathematical fictionalist shows that in fact his model-theoretic reasoning does a lot more work than it is supposed to do: it is in fact vital to the understanding of logical possibility, a straightforward *reductio* of his original claim. Either way then, there appears to be no way of *understanding* these subscripted modal operators without ontological commitment to models. And since these subscripted modal operators are necessary for a complete exposition of the central notion of logical possibility, mathematical fictionalism faces an irreducible commitment to the existence of abstract mathematical objects.

4
On the Nature and Norms of Acceptance and Belief

4.1 Another perspective on scientific attitudes

Constructive empiricism – as articulated and defended by van Fraassen – is intended as but a component of a broader picture of our epistemic lives; it is a conservative assessment of the aim of science motivated not so much by scepticism at our ability to acquire knowledge about the unobservable as it is by a particular orientation towards explanation, a visceral disinclination for speculative metaphysics and a permissive conception of rationality that demands no greater license for a philosophical position than its logical consistency. And initially, this voluntarist framework appeared to provide the ideal habitat for van Fraassen's empiricism. The most pressing objection facing a position that claims that the acceptance of a scientific theory involves as belief only that it is empirically adequate is that, since there appears to be no principled epistemological distinction between what our theories say about the very large (but extraordinarily distant) and what they say about the very small (but routinely detected), any justification we may have for believing a theory to be empirically adequate will also be justification for believing that theory to be true. But the constructive empiricist is not offering us an argument concerning the warrant of our various scientific beliefs – he is offering us an alternative conception of the scientific enterprise against a background epistemological framework that rejects all questions of justification as ultimately an expression of taste.

We have however seen two problems with this picture. The first is that epistemic voluntarism – essentially *carte blanche* to avoid answering difficult epistemological questions – is far from being an intuitive philosophical position, and van Fraassen's arguments in its support leave a lot to be desired. The doxastic constraint of diachronic probabilistic coherence certainly complements the voluntaristic perspective; but since diachronic coherence is only compelling for those who already endorse a voluntaristic attitude towards epistemic judgements, it hardly constitutes an independently compelling case for van Fraassen's minimalist conception of rationality. With respect to our inferential practices, van Fraassen does raise pertinent sceptical difficulties concerning our methods of ampliative inference; yet these arguments are too corrosive to motivate epistemic voluntarism, since in order to undermine those rule-based practices that are his target, van Fraassen must invoke a sufficiently radical scepticism that leaves his inferentially permissive alternative equally groundless. And while the prospect of a stance-based conception of empiricism is an interesting one, van Fraassen has yet to show any conclusive advantage it may have over its doctrinal ancestor, or to provide sufficient detail to show how it doesn't simply collapse into full-blown relativism.

Secondly, and more importantly, we have also seen how epistemic voluntarism is utterly unable to resolve those problems concerning the internal coherence of constructive empiricism. The basic difficulty is that in order to maintain his crucial distinction between the observable and the unobservable (as well as maintaining a philosophically satisfying account of modality and mathematics), the constructive empiricist appears to be committed to some of those very beliefs about unobservable phenomena (nomological possibility, the existence of abstract mathematical objects) that such an account was meant to render illegitimate. The general response to this problem advocated by van Fraassen has been to defend an impoverished – but internally coherent – notion of observability (modality, mathematics), and to try and make up the doxastic deficit with a series of arbitrary stipulations that are justified only insofar as they are to be articulated within the context of a voluntaristic framework in which substantial epistemological questions are moot. Unfortunately, however, the problem of internal coherence is more pervasive than van Fraassen gives it credit: since even the notion of logical consistency is one for which the constructive empiricist struggles to provide an internally coherent account, it is clearly unacceptable to appeal to an epistemological framework that presupposes such a notion (even if it presupposes very little else) in order to justify any attempt to salvage one's position. Epistemic voluntarism cannot therefore rectify the constructive empiricist's account of abstract mathematical objects; since this account underlies his views on nomological possibility, epistemic voluntarism cannot therefore rectify the constructive empiricist's account of modality; and since *this* account underlies his views on empirical adequacy, epistemic voluntarism also cannot rectify the constructive empiricist's fundamental distinction between observable and unobservable phenomena. In short, van Fraassen's proposed epistemic framework is both poorly motivated and utterly incapable of resolving the most fundamental difficulties with constructive empiricism. All of this motivates exploring an alternative perspective upon our (empiricist) attitudes to science.

My proposal is a very simple one. In addition to the distinction between observable and unobservable phenomena, constructive empiricism also depends upon an equally crucial distinction between accepting the claims of a successful scientific theory and believing those claims. Constructive empiricism is the view that the acceptance of a scientific theory involves as belief only that the theory in question is empirically adequate; however, the claims made by an accepted scientific theory with regard to the unobservable phenomena are not thereby abandoned, ignored or reduced to further claims about observable phenomena – they are an ineliminable (and irreducible) component of the success of the theory as a whole, to be used in making predictions and offering explanations in just the same way as the claims the theory makes about observable phenomena. It follows then that when the constructive empiricist accepts a scientific theory, he accepts *every* claim made by that theory; yet since he only believes a specified subset of those claims that he accepts, the two attitudes must be distinct. The idea then is this: if the basic problem facing the internal coherence of constructive empiricism is that in order to endorse the position one is committed to certain claims that one cannot also believe, but if it is also part and parcel of the constructive empiricist's position that there are a variety of different and epistemologically substantive attitudes that one can hold towards the claims of our scientific theories, then maybe all the constructive empiricist needs to do is to *accept* these problematic claims about unobservability, nomological possibility and the existence of abstract mathematical objects. In contrast then to van Fraassen's strategy, which turns upon lumbering the constructive empiricist with a problematic distinction between the different standards of *rationality* that one can appeal to in the evaluation of a philosophical position, the current proposal is to simply elaborate upon a distinction between the different *attitudes* that one can hold towards a proposition or theory that is already integral to the view under discussion.

Of course, in order to put the distinction between acceptance and belief to work in support of the distinction between observable and unobservable phenomena, one must first show that there really is a philosophically satisfactory distinction between these two attitudes to which one can appeal; and as I remarked briefly in Chapter 1, while the literature surrounding the distinction between acceptance and belief is far less extensive than the literature concerning the distinction between observable and unobservable phenomena, it is if anything even more critical. The worry here is that however the constructive empiricist is going to characterise acceptance in contrast to belief, he must satisfy two competing criteria: on the one hand, acceptance must be sufficiently robust such that the constructive empiricist can make predictions, offer explanations and regiment his scientific practice on the basis of those claims that he merely accepts (or else face an impossibly impoverished account of science that effectively rejects the unobservable content of our scientific theories); on the other hand, acceptance must not be so robust that it seems to be belief under a different name (in which case constructive empiricism would collapse into a merely verbal alternative to scientific realism). And for many commentators, this difficulty is insurmountable.

In the first half of this chapter then, I consider the various objections to the constructive empiricist's purported distinction between acceptance and belief, and sketch some proposals for developing this aspect of his position. In particular, considerations over the voluntariness of the two attitudes (voluntary in the sense that one can choose what one accepts but not what one believes, and not to be confused with what is in effect van Fraassen's relativism about rationality), and the way in which intentional states are to be individuated, give us a sense in which the two attitudes differ; and an exploration of some of the difficulties facing an adequate epistemology of science in the face of the various contradictions that permeate scientific practice provides further argumentation that something along the lines of the distinction between acceptance and belief is philosophically desirable, quite independently of the issues surrounding constructive empiricism. In the second half of this chapter, I explore how this distinction can be put to work in the articulation and defence of constructive empiricism: that by focusing upon the different attitudes involved in our understanding of scientific practice, rather than the different standards of rationality, the constructive empiricist can successfully respond to the objections raised by Musgrave, Ladyman and Rosen.

In some ways this may appear to be a retrogressive step – what initially motivated our discussion of epistemic voluntarism was the use made of it by van Fraassen in dissolving those issues relating to the justification of the constructive empiricist's views regarding the aim of science; having now rejected this conception of rationality outright, it seems that the justificatory challenges encountered in Chapter 1 must thereby loom back into view. But this need not be the case – that epistemic voluntarism is unable to resolve the problem of internal coherence is not an argument that the constructive empiricist must therefore reject this minimalist conception of rationality altogether; it is an argument that the constructive empiricist must, at the very least, *augment* his minimalist conception of rationality along the lines sketched in this chapter. Everything that I propose here is perfectly consistent with retaining a voluntarist epistemology, and consequently dismissing the justificatory challenge on the grounds of its background epistemological presuppositions. That being said however, we should also remember that the actual justificatory challenges we encountered in Chapter 1 – as put forward by philosophers such as Churchland and Hacking – were rather less compelling than their advocates would have us believe, irrespective of any concerns regarding our background epistemological framework. Having such justificatory challenges loom back into view will not be cause for constructive empiricist alarm. Thus while my proposal here is certainly *consistent* with epistemic voluntarism, we may well wonder – in the absence of any compelling motivation for retaining van Fraassen's minimalist conception of rationality – whether or not it would be better taken as a *replacement* for epistemic voluntarism.

In any case (and perhaps more importantly) my proposed shift of emphasis from the different standards of rationality that one can appeal to in the evaluation of a philosophical position, to the different attitudes that one can hold towards a proposition or theory, has the added advantage of opening up some interesting new ground in the philosophy of science. My overall aim in this chapter therefore is to sketch an internally coherent articulation of constructive empiricism, liberated from van Fraassen's broader epistemological perspective, and which thereby offers a fresh perspective upon the epistemological dimensions of the scientific realism debate.

4.2   Deciding to accept, deciding to believe

4.2.1   A lot of fuss about functionalism

Although most of the criticisms of constructive empiricism concern the distinction between observable and unobservable phenomena, many critics have also objected to the distinction between acceptance and belief. Yet if this second line of attack has been perhaps less prominent in the literature, it has certainly been more *consistent* – whereas criticisms of the former distinction have variously concerned the relative justification of instrumentally aided observations, the counterfactual stability of various unobservable properties and relations, or indeed the very plausibility of selectively abductive inference, criticisms of the latter distinction have by contrast shown remarkable uniformity in their complaint that since so much is packed into the notion of acceptance, one wonders where this attitude is supposed to stop and the putatively contrasting attitude of belief begins. This in part explains the paucity of literature on this issue. More important, however, is the impression that, if criticism of the distinction between acceptance and belief has been perhaps less prominent in the literature, it has without doubt been more *conclusive* – not only is the problem perfectly clear, it is even clearer that it cannot be resolved. It is frequently noted that once one takes into account the substantial epistemological commitment that is constitutive of the constructive empiricist's acceptance of a scientific theory – reliance upon its predictions, pursuing a research programme on the basis of its theoretical framework, immersion within the theory for the purposes of providing explanations and so on and so forth – there appears to be nothing left to be contrasted with the supposedly distinct attitude of belief. In an early and paradigmatic expression of this worry, Melchert puts the problem as follows:

> The question to be asked is obvious. If Jones, let us say, accepts a certain theory, relies on it to make predictions, uses it in solving practical or technical problems, is 'conceptually immersed' in the theory, asks and answers relevant questions using the language of the theory, confronts new phenomena using the theory's resources, and seeks explanations in terms supplied by the theory – in what sense does Jones not *believe* that theory?
>
> (1985: 214)

Such rhetorical touches are easily multiplied; but it is perhaps, as Melchert notes, Peirce who illustrated the underlying worry best when he wrote:

> Do you call it *doubting* to write down on a piece of paper that you doubt? If so, doubt has nothing to do with any serious business.
>
> (1934: 278)

Frequently however this is often as far as the argument is developed: for all the enumeration of the myriad characterisations that must be packed into the constructive empiricist's notion of acceptance, and for all the rhetorical flourishes and incredulous stares, one is often left with the fairly bold assertion that since there is a lot involved in acceptance, it must be the same as belief after all, the constructive empiricist's insistence notwithstanding. It is the inevitable flipside to offering a conclusion that provokes such ready assent that all too often, too little attention is paid to what exactly the *argument* is supposed to be.[1]

In order to proceed therefore, we must attempt to identify the underlying reasoning behind this dissatisfaction with the constructive empiricist's distinction between acceptance and belief; and it is to this end that I propose to focus our attention upon Paul Horwich's (1991) presentation of the objection. Horwich develops his argument from the perspective of the philosophy of mind, rather than from the philosophy of science *per se*. Nevertheless, he manages to make explicit the concerns shared by many of the constructive empiricist's critics; and, moreover, he manages to present them in an initially compelling manner. For these reasons then, my discussion in this chapter will be directed exclusively as Horwich's presentation of the problem – that of the many criticisms of the constructive empiricist's distinction between acceptance and belief, Horwich's argument is one of the few to offer independent reasoning against its coherence (rather than dismissing the distinction on the grounds of an antecedent rejection of constructive empiricism); that of these many criticisms of the constructive empiricist's distinction, Horwich's argument is one of the few to satisfactorily diagnose the underlying dissatisfaction common to all such criticisms, and to present it in an initially compelling line of objection; and that most importantly, for both of these two reasons outlined above, a successful response to Horwich will therefore *a fortiori* constitute a successful response to any other variation of the problem.[2]

Horwich begins his argument by noting that both acceptance and belief are some kind of mental state, and then goes on to make the reasonable assumption that *qua* mental state, both acceptance and belief should be characterised in terms of their functional role. Among his various writings, van Fraassen himself gives us a good guide as to the functional characterisation of acceptance, much of which we have already encountered: it involves making inferences with the theories that we accept, using them in predictions, 'confronting any future phenomena by means of the conceptual resources of [that which we accept] ... it is exhibited in the person's assumption of the role of explainer, in his willingness to answer questions *ex cathedra'* (van Fraassen, 1980: 12). To make the same point in explicitly functionalist terms, to accept something is to be in a state caused by certain observations (the apparent empirical adequacy of the theory in question), which enters into certain inferences and deliberations, and which in turn generates certain predictions and utterances (the offering of an explanation being a case in point).

But if that is the case, argues Horwich, accepting a theory sounds a lot like *believing* a theory:

> If we tried to formulate a psychological theory of the nature of belief, it would be plausible to treat beliefs as states with a particular kind of causal role. This would consist in such features as generating certain predictions, prompting certain utterances, being caused by certain observations, entering in characteristic ways into inferential relations, playing a certain part in deliberation, and so on. But that is to define belief in exactly the way [van Fraassen] characterises acceptance.
>
> (1991: 3)

Horwich concludes that since acceptance and belief are characterised by the same causal role, they must in fact be the same mental state. Consequently, the constructive empiricist's distinction between accepting a scientific theory – an epistemological commitment that need only be accompanied by belief towards certain parts of that theory – and believing a scientific theory *simpliciter* collapses, and with it constructive empiricism in general.

In this chapter I intend to provide a straight solution to Horwich's objection, that is, to demonstrate in detail how the two attitudes of acceptance and belief *do* in fact differ from one another in terms of their functional characterisation. But before turning to this positive project, it is important to elaborate upon some of the explicit presuppositions of Horwich's argument – and by extension, to help make explicit those less articulated presuppositions that lie behind the general misgivings facing the constructive empiricist's distinction. As I noted above, the root anxiety seems to be that once one enumerates all of the myriad applications involved in the constructive empiricist's notion of acceptance, there remains little else to be contrasted with the putatively distinct attitude of belief; whether or not this is in fact true (and I will later reject this claim), it is clear that such an assertion could only have argumentative force if one also assumes that a mental state is *exhausted* by its causal role – Horwich's presupposition of a functionalist philosophy of mind therefore is not an idiosyncrasy of his own presentation of the problem, but the explicit manifestation of an essential ingredient of any such criticism of the distinction. It follows then that if we were to argue – as many have – that mental states cannot be completely identified in terms of their functional role since there will always remain some *phenomeno-logical* residue that escapes characterisation in terms of causes and effect (see, e.g., Jackson, 1986 – although he has of course since abandoned this point of view), then the entire case against the constructive empiri-cist's distinction would collapse: for if such a phenomenological thesis were true, then the functional equivalence of acceptance and belief would not establish their identity, since each attitude may in fact be accompanied by a distinctive qualitative aspect. Moreover, even if we could exhaustively specify acceptance and belief in terms of their func-tional role, and even if these functional characterisations turned out to be identical, we could only conclude that the two attitudes were one and the same if we took the further step of identifying a mental state with its functional role rather than with whatever it is that *realises* this functional role: acceptance and belief may be functionally identical, but as a matter of empirical fact be realised by distinct brain states. In short, what Horwich's presentation makes clear is that the various misgivings facing the constructive empiricist's distinction don't just assume a func-tionalist theory of mind, but assume a *functionalist state identity theory* of the mind; and this is far from uncontroversial (see, e.g., Lewis, 1980).

The second underlying presupposition of the near ubiquitous dis-missal of the distinction between acceptance and belief – and here we begin to make contact with my positive project – appears to be an unre-alistically broad characterisation of the functional role of belief. In the next section I will begin to elaborate upon the various respects in which the myriad applications brought to bear in our characterisation of belief may in fact be better thought of as a characterisation of acceptance; here it is sufficient simply to prepare the ground for my purposed division of doxastic labour by noting some further weaknesses in the argument that we have uncovered. To take a particularly good example: as many of the constructive empiricist's critics have intimated – and as Horwich makes explicit – one of the most important sets of candidates for the func-tional characterisation of belief are the various utterances, predictions and explanations that we are supposed to make as a result of having such a belief. Yet not only is it far from clear how possessing the dis-position to make the relevant utterances could be either necessary or sufficient for possessing the belief in question, it is moreover perfectly clear that the domain of our utterances and predictions cannot be the exclusive right of our beliefs. As Williams (1973: 140–141) has noted, many of our utterances are insincere – insofar as they are utterances of propositions that we do not believe – and conversely, many of those propositions that we do believe live out their days never having been asserted.[3] But if this is the case, then not only do we need to reject the close connection between utterance and belief that underlies Horwich's argument, we need also to reject the supposed monopoly of belief over the domain of our utterances and predictions and open up the conceptual space for something like the constructive empiricist's distinction between acceptance and belief. If what makes an assertion sincere is that we also believe it, then being disposed merely to make such an assertion cannot be a functional characterisation of belief since then it would be impossible to make an *insincere* assertion. Simply being disposed to make an assertion – to answer questions *ex cathedra*, as it were – must therefore be the province of another mental state, one clearly distinct from the notion of belief. Moreover, if we agree with Williams that it is a *necessary* feature of our notion of belief that we have both 'the possibility of deliberate reticence, not saying what I believe, and if insincerity, saying something other than what I believe' (1973: 147), then it seems that any adequate characterisation of belief will necessarily entail a contrast with whatever mental state is responsible for our insincere assertions.

The underlying discontent with the constructive empiricist's distinction is therefore not only based upon an extremely specific, and in many ways highly contentious, philosophy of mind; it is also based upon an unrealistically broad conception of the notion of belief, one that threatens to obscure a number of salient features of our epistemic and/or doxastic practices in its unreflective appropriation of all and any causal manifestation of our mental lives as symptomatic of belief. The weakness of these two presuppositions undermines much of the case against the constructive empiricist; the burden of proof still lies with those, for whom I have appointed Horwich spokesman, to provide any kind of satisfactory argument against the distinction between acceptance and belief.

Nevertheless, it is still important to provide a positive characterisation of the differing functional roles of acceptance and belief – especially if, as I have suggested above, this distinction is to be put to work in the general articulation and defence of constructive empiricism. My basic strategy therefore will be to argue that acceptance and belief are to be understood as two *entirely distinct* attitudes. This will of course make explicit how the two attitudes functionally diverge, and thus secure the constructive empiricist's desired conclusion; more importantly, however, it is only through distinguishing the two attitudes in this way that we can hope to put the distinction between acceptance and belief to work in defence of the constructive empiricist's other crucial distinction between observable and unobservable phenomena. If acceptance and belief were not entirely distinct, then one attitude would have to be characterised in terms of the other. For example, one natural suggestion is to define acceptance as a kind of suitably restricted belief – the belief that a scientific theory is empirically adequate, say, rather than the unrestricted belief that it is true. But with respect to our more ambitious concern, the problem with such a characterisation is obvious: if the acceptance of a scientific theory is defined as the belief that the observable consequences of that theory are true, then one must appeal to the distinction between observable and unobservable phenomena in order to ground the notion of acceptance; one cannot then also appeal to the notion of acceptance in order to ground the distinction between observable and unobservable phenomena. The prospects are no better in the other direction, that of attempting to define belief in terms of acceptance. Paul Teller (2001a) has made just such a suggestion: according to him, we should understand acceptance as the fundamental attitude, and understand belief as constituting a special, limiting case. Specifically, he argues that acceptance is to be understood as the attitude of treating a proposition or theory as if it were true under certain conditions (that is to say, only for particular purposes, or only over restricted domains of application), whereas belief is the attitude of treating a proposition or theory as if it were true without such restrictions – we can think of belief as the ideal limit to which acceptance approaches.

There is much to recommend in Teller's suggestion. On the one hand, such a characterisation of the distinction between acceptance and belief allows us to accommodate the intuition that both attitudes are in fact very closely related (a fact upon which Horwich's argument attempts to capitalise); on the other hand, it also allows us to accommodate a broad range of distinctions within the notion of acceptance itself – such an attitude can come in degrees, and the way in which one might accept an idealisation in physics need not be the same as the way in which one accepts a methodological principle in biology, for example. Finally, and with respect to our present purposes, it also looks as if Teller's proposal provides the constructive empiricist with an extremely straightforward response to Horwich: acceptance and belief must be distinct attitudes insofar as they functionally diverge over their respective *scopes*.

Nevertheless, an adequate defence of constructive empiricism can no more allow belief to be defined in terms of acceptance than it can allow acceptance to be defined in terms of belief. By far the most substantial difficulty in this respect is that Teller (2001a: 143–145) actually intends his characterisation of acceptance and belief to constitute a *refutation* of constructive empiricism: he argues not only that belief is to be understood as a limiting case of acceptance, but moreover that belief – once so construed – is more or less impossible, thus undermining any philosophical proposal that rests upon employing a substantial distinction between acceptance and belief. According to Teller, all scientific theories are only ever treated as if they were true under very specific conditions; the sort of universal validity implicit in the notion of belief is nothing more than a philosopher's pipe dream. This view is in turn based upon Teller's (2001b) conception of what a scientific theory *is*; following such philosophers as Giere (1988; 1999) and Cartwright (1983; 1999), Teller argues that we should understand a scientific theory to consist of a set of models, and that we should understand a model by analogy to a map – that is, accurate to varying degrees *relative to a set of goals or interests*. The London Tube Map is accurate with respect to the topological relationships between the different stations, for example, but not with respect to the actual colour of the rails linking them together. Consequently, since all scientific theories are only ever regarded as true with respect to certain purposes, the constructive empiricist's distinction between acceptance and belief collapses, although from the opposite direction than argued for by Horwich: according to Teller, it is belief that collapses into acceptance.[4]

It may of course be possible to adopt Teller's characterisation of the distinction between acceptance and belief without endorsing his views regarding the restricted application of a scientific theory. But there is another difficulty with this account – one with respect to our more ambitious aim of not only defending the constructive empiricist's distinction between acceptance and belief, but of subsequently using this distinction in response to the difficulties raised by Musgrave, Ladyman and Rosen. The problem is that, according to Teller, to accept a proposition or theory is to treat it as if it were true under very specific circumstances, whereas to believe a proposition or theory is to treat it as if it were true under every possible circumstance. For the constructive empiricist then, it follows that those putatively problematic propositions concerning the unobservability of unobservable phenomena (nomological possibility, the existence of abstract mathematical objects) will necessarily fall into the latter camp: since the distinction between observable and unobservable phenomena (and *mutatis mutandis* for the modal and mathematical manifestations of this distinction) in part *determines* which attitude the constructive empiricist is to take towards the other consequences of his accepted scientific theories, the propositions that constitute this distinction must therefore be *presupposed* by the constructive empiricist in adopting any other attitude towards any other proposition whatsoever. Those putatively problematic propositions in which we are interested must therefore be treated as if they were true under every possible circumstance since they provide the framework in which the constructive empiricist is to determine which propositions are to be treated as if they were true under every possible circumstance. Such propositions, simply on the grounds of their universal scope, are therefore clear candidates for what Teller wishes to classify as belief. The upshot of all this then is that if the constructive empiricist agrees with Teller that our notion of belief is to be defined as the limiting case of our notion of acceptance, and thus if the constructive empiricist endorses the view that he must *believe* those putatively problematic propositions concerning the unobservability of unobservable phenomena, the similarity ordering of possible worlds and the existence of abstract mathematical objects, then there is no hope of responding to the criticisms raised concerning the internal coherence of his philosophy of science – for these criticisms are precisely the claim that the constructive empiricist *cannot* believe such propositions.

If acceptance is defined in terms of belief, then the distinction between acceptance and belief cannot be used in defence of the internal coherence of constructive empiricism since it will in fact *presuppose* that internal coherence; and if belief is defined in terms of acceptance, the distinction between acceptance and belief cannot be used in defence of the internal coherence of constructive empiricism since it will simply *endorse* what is taken to be the root of the problem. Any satisfactory defence of the distinction between acceptance and belief must therefore consider the two attitudes as being entirely distinct.

As it transpires, however, there is no strong connection between these two attitudes anyway: one can accept what one does not believe, and one can believe what one does not accept. The first possibility, of acceptance without belief, is in my view the correct way to describe the sort of situation where one uses a scientific theory for various pragmatic reasons even though one cannot bring oneself to believe it. This would be the case when scientists employ manifestly false idealisations – such as frictionless planes and perfectly elastic collisions – for the purposes of simplifying predictions, or perhaps using Newtonian mechanics for systems with sufficiently small velocities such that any (relativistic) inaccuracies will be negligible. The converse situation, of not accepting what one believes, is perhaps a little less obvious. But in my view this would be the correct way to describe the sort of situation where one is forced to ignore some of one's beliefs when calculating the most appropriate course of action: a conscientious scientist, say, who must admit that he simply lacks sufficient evidence for his own deeply held convictions; or for a more homely example, a juror who bases his verdict upon the evidence presented despite any additional beliefs he may have concerning the accused.

Of course, considerations such as these are only convincing with respect to a more detailed account of the distinction between acceptance and belief. But what I want to make clear is that not only does the case against the constructive empiricist's distinction rest upon some highly questionable assumptions, but moreover that the case in favour has some intuitive plausibility – not only must a satisfactory functional characterisation of belief leave room for the other mental states with which it must be contrasted, but so too do the range of cases for which our mental states are to provide the interpretative framework demand a distinction along the lines proposed by the constructive empiricist.

4.2.2 The individuation of intentional states

My preliminary orientation to the distinction between acceptance and belief has been to emphasise that the two attitudes must be considered as entirely distinct, since if one attitude were defined in terms of the other there would be no prospect of subsequently using such a distinction in the articulation and defence of constructive empiricism. Although he rarely discusses the issue directly, van Fraassen's view also seems to be that the two attitudes must be entirely distinct: in an early defence of the distinction, van Fraassen (1985: 276–281) argues that since there are reasons one might have to accept a proposition or theory that are not thereby reasons for believing that proposition or theory (and *vice versa*), the two attitudes must be distinct. More specifically, van Fraassen argues that one comes to accept a scientific theory on what we might call largely *pragmatic* grounds: for example, that a particular scientific theory is easy to use, simple to understand or, indeed, merely coherent with the rest of one's scientific worldview are all motivations for accepting that theory. By contrast, van Fraassen argues that we come to believe a proposition or theory on largely *credential* grounds: how likely it seems to us that the proposition or theory is true. Crucially, these considerations can come apart. To take a simple example: the more claims that a scientific theory makes about the world, the more useful that theory is (on account of its wider domain of application), and thus the more reason we would have to accept it; yet the more claims that a scientific theory makes about the world, the more chance it has of saying something false (on account of its greater number of potential falsifiers), and thus the less reason we would have to believe it. As a limiting case then, a tautology such as (p ∨ ¬p) would be *maximally believable* in the sense that there would be no way for it to be false; yet it can be *minimally acceptable* in the sense that, since it tells us nothing about the way the world is, we would have absolutely no motivation to include it among our action-guiding principles.

This then is a response that can directly engage with Horwich's argument: for if van Fraassen is right, and the causes of belief are to be differentiated from the causes of acceptance, then by Horwich's own functionalist standards the two attitudes must be distinct. Moreover, we can clearly see how van Fraassen's response accommodates both of our initial intuitions concerning the intelligibility of the distinction between acceptance and belief: to accept a proposition or theory even though one does not believe it (in the sense, for example, of utilising a manifestly false idealisation) is to be motivated by largely *pragmatic* considerations; and conversely, to believe a proposition or theory even though one does not accept it (in the sense, for example, of the juror with extra-evidential opinions concerning the accused) is to distinguish between what one finds *credible* and what one takes to be appropriate. Thus van Fraassen's characterisation easily maps onto the sorts of cases that we assume it must. Furthermore, the contrast between pragmatic and credential considerations also accommodates the intuition that one cannot understand the functional characterisation of one attitude without the other: an insincere utterance, to return to Williams' example, is clearly a pragmatic activity with no necessary connection to what an agent finds credible; in contrast, those beliefs that are never uttered attest to the fact that it is not always in an agent's interests to assert what they do find credible. Horwich (1991: 7–8) however simply dismisses this line of thought. According to Horwich, van Fraassen's distinction between the pragmatic and the credential underdetermines the case at hand – while our mental lives may indeed be governed by both instrumental and evidential considerations as van Fraassen suggests, this in itself is insufficient to establish that we are dealing with two distinct attitudes; equally plausible is the view that there is only one attitude under consideration, one which happens to have both pragmatic and credential elements. In Horwich's view then, all that van Fraassen has shown is that our decision to *believe* a proposition or theory will depend upon both its likeliness and its loveliness, considerations that can pull apart for sure; and none of this is sufficient to establish a principled distinction between the two contrasting attitudes of acceptance and belief.[5]

Given the various weaknesses we identified in Horwich's actual case against the constructive empiricist's distinction (the highly specific philosophical theory of mind, the extremely crude characterisation of belief), and given the initial plausibility in favour of such a distinction (a more nuanced characterisation of belief, the range of cases that require a multi-attitudinal interpretative framework), the burden of proof here must lie squarely with Horwich and those critics for whom he speaks. The point is further compounded when we note that in some fields outside of the philosophy of science, something like the constructive empiricist's distinction between acceptance and belief almost passes as common currency. To take just one example: in the philosophy of the social sciences there remains an ongoing debate as to whether collective intentionality – the intentional states that one might wish to attribute to a group, over and above those intentional states that one might wish to attribute to the individuals that make up that group – is best conceived of in terms of acceptance or belief. To relate this explicitly to the foregoing discussion, one of the issues here is that while one may wish to attribute certain action-guiding states to a collective over and above those action-guiding states that one may wish to attribute to its individual members (one may wish to say that a committee or corporation is following a particular policy even in those situations where it is clear that no single member of that committee is following that policy), one may still balk at attributing *beliefs* to that collective on the grounds that this implies some kind of representational capacity of which only 'real' intentional agents are capable. Such reasoning is of course highly controversial; the point however remains that most parties to the dispute at least agree that there is a meaningful distinction between those intentional attributions that are primarily concerned with pragmatic considerations and those intentional attributions that are primarily concerned with credential considerations – exactly the sort of distinction that the constructive empiricist's notions of acceptance and belief are meant to capture – in terms of which the debate over collective intentionality can be framed.[6] So while Horwich is right to note how van Fraassen's considerations underdetermine the matter in hand, it seems that the burden of proof must remain with the constructive empiricist's critics.

More positively though, van Fraassen's basic contention – that acceptance is essentially pragmatic, while belief is essentially credential – can be further elaborated upon so as to conclusively dispel Horwich's misgivings. The central contrast between the two attitudes that I wish to develop concerns the issue of voluntarism. This of course immediately raises some terminological confusion, since the notion of voluntarism at issue in the demarcation of acceptance and belief is not the same as the notion of voluntarism at issue in van Fraassen's conception of rationality. In our present context, we are interested in whether or not an attitude is voluntary in the sense that an agent can sit down and decide whether or not to hold it towards a particular proposition or theory. Expanding then upon van Fraassen's initial characterisation, it is clear that acceptance is a voluntary attitude in this sense: as an essentially pragmatic attitude, there must therefore be a decision to accept a particular proposition or theory based primarily upon the perceived utility of so doing. Acceptance is always a matter of choice: for no matter how well a given scientific theory accommodates the relevant data, meets our explanatory needs, coheres with our other accepted scientific theories or even appeases our aesthetic sensibilities, one can always choose not to accept it. Even in those cases where it may appear that we have no choice in the matter – for instance, when the theory in question is the only one available for dealing with a particular range of phenomena – it would still be wrong to think of ourselves as confronted with a case of involuntary acceptance: for no matter how great one's practical need to employ a particular theory, one could always choose not to engage with the issue at all. In sum, since acceptance is an attitude primarily concerned with what an agent *does* – which premises he chooses to reason with, which assertions and predictions he chooses to make or according to which authority he chooses to speak *ex cathedra* – it is thereby essentially voluntary.

By contrast, I take belief to be essentially involuntary in the sense that one cannot simply choose what one believes. Consider for example our perceptual beliefs: one of the common assumptions concerning this class of beliefs is that an agent will come to believe what they do purely on the basis of a (presumably causal) relationship that exists between the environment and their perceptual faculties – there is no intermediary stage where one can consider one's options and exercise a degree of choice. This however is not the sense in which van Fraassen discusses the voluntarism of belief. When van Fraassen speaks of a voluntarist epistemology, he is not suggesting that we can decide what we wish to believe (see Psillos, 2007: 140–141; this is made explicit in van Fraassen, 2007: 351). On van Fraassen's view, it is certainly the case that different epistemic agents (empiricists, speculative metaphysicians) will come to believe different propositions, often in the face of the exact same evidence; and since there are no standards for adjudicating such disagreement, beyond of course the minimal constraints of logical consistency and probabilistic coherence, what an agent believes can therefore be said to be voluntary insofar as there is no sense in which it can be obligatory. But all of this should be seen as the manifestation of different epistemological commitments, rather than as the outcome of some conscious deliberation on the part of our epistemic agents. It is helpful to think of the situation not so much as an epistemic agent *adopting* an epistemic stance (which would imply voluntarism in the sense of making a decision), as it being appropriate to *attribute* different epistemic stances to an agent on the basis of their manifest commitments (in which case, any notion of voluntarism would be confined merely to the sense in which there are no objective standards to judge the relative merits of these various epistemological frameworks). So, according to this distinction, belief can be voluntary in van Fraassen's sense that there are no overridding normative constraints – no objectively right answers – as to what an agent is rationally obliged to believe; while at the same time belief can be involuntary in my sense that an agent cannot sit down and come to a decision as to what he will in fact come to believe. Thus even those who endorse van Fraassen's permissive conception of rationality could similarly endorse my proposed demarcation between acceptance and belief; and those who reject my characterisation of acceptance and belief need not therefore find themselves lumbered with a minimalist epistemology.

To return then to the matter at hand: acceptance, I have argued, is a voluntary attitude on the grounds that it is essentially pragmatic and concerned with what an agent *does*; belief, I contend, is an involuntary attitude on the grounds that it is essentially credential and concerned merely with *how things appear to be* for a particular agent. Yet while the voluntarism of acceptance seems fairly intuitive, the involuntarism of belief is a matter of greater controversy. The most famous case made in favour of this thesis – and the starting point for our discussion – is due to Williams (1973), who has argued that it is *conceptually necessary* that one cannot choose what one believes. To take his oft-quoted argument:

> One reason [that one cannot believe at will] is connected with the characteristic of beliefs that they aim at truth. If I could acquire a belief at will, I could acquire it whether it was true or not. If in full

consciousness I could will to acquire a 'belief' irrespective of its truth, it is unclear that before the event I could seriously think of it as a belief, i.e., as something purporting to represent reality. At the very least, there must be a restriction on what is the case after the event; since I could not then, in full consciousness, regard this as a belief of mine, i.e., as something I take to be true, and also know that I acquired it at will. With regard to no belief could I know... that I had acquired it at will. But if I can acquire beliefs at will, I must know I am able to do this; and could I know that I was capable of this feat, if with regard to every feat of this kind I had performed I necessarily had to believe that it had not taken place?

(1973: 148)

To believe a proposition or theory is to believe that the proposition or theory in question is true. But if one could believe at will, one could choose to believe a proposition or theory irrespective of its truth; and moreover, one would also know that this is what one had done. The problem however is that in order for an agent to consider a voluntary belief *as* a belief – that is, consider it as true – he must therefore not know that the belief had been acquired at will. But how then could someone acquire a belief at will if, in order to do so, he must both know that this is what he was doing *and* not know that this is what he was doing?

The problem with Williams' argument however is clear – it bases the impossibility of believing at will upon the impossibility of *knowing* that one had acquired a belief at will, and is therefore vulnerable to the sort of counterexample where the purported doxastic voluntarist *forgets* (or indeed, is never even aware of the fact) that this is what he has done. Bennett (1990), for example, has offered the possibility of a tribe of 'Credamites', individuals who whenever they choose to believe something (an admittedly rare occurrence, constrained by the availability of at least *some* evidence in favour of their belief), they instantly forget that their belief had been so acquired. It follows then that each Credamite 'knows that he sometimes wills himself to believe something, even though it is never true that he *now* has a belief which he *now* remembers having willed himself to acquire' (Bennett, 1990: 93). The Credamites therefore can believe at will, know that they can believe at will, yet never find themselves in Williams' predicament of holding inconsistent opinions about their own mental states.

Nevertheless, many philosophers – Bennett included – take Williams' thesis to be essentially correct, and several reformulations of his argument that do not rest upon *knowing* that one had acquired a belief voluntarily have been proposed (although none has generated universal consent).[7] This remains a substantial philosophical debate in its own right, and beyond the scope of this book to resolve. Fortunately, however, for our present purposes, Williams' original argument furnishes us with all the material that we need. For our main concern is with the distinction between the constructive empiricist's notions of acceptance and belief; and while Williams' argument may fall short of establishing the conceptual impossibility of a voluntary belief *tout court*, it does establish enough to draw a principled contrast between the two attitudes. At the very least, we must concede that it is conceptually impossible for an agent to *consciously hold* a voluntary belief – that is, while we may have to concede that some beliefs can be acquired at will, we can certainly deny that such an eventuality can be recognised as such by any self-reflective agent. One can believe at will (perhaps); but one cannot recognise oneself as believing at will. Contrast this with the case of acceptance: not only can one accept at will, but since such a decision is based upon pragmatic considerations which may or may not vary as the circumstances alter, it is clear that one is always conscious of this fact too. One can accept at will (definitely); and one will always recognise oneself as accepting at will. Indeed, if the attitude of acceptance were not open to such transparent self-recognition, it would be difficult to see how it could be based primarily upon pragmatic considerations – if one were not aware that one had accepted a proposition or theory on the basis of the utility of so doing, then one would not be aware that one could *cease* to accept that proposition or theory when a better alternative presents itself, and one would therefore not be acting out of primarily pragmatic considerations.

Although I take Williams' argument for – at the very least – the conceptual *oddity* of believing at will to be highly compelling, it must be conceded that it does not establish the conclusion for which we are after: that the constructive empiricist's distinction between acceptance and belief can be conclusively grounded upon the fact that the former attitude is completely voluntary, whereas the latter attitude is completely involuntary. We can get part of the way there by noting that acceptance can be *regarded* as completely voluntary – in contrast to belief which cannot be so regarded – by the agent in question; although one can easily imagine a trenchant critic like Horwich simply retorting that such considerations can only show us something about an agent's (no doubt deeply confused) understanding of his own mental states, and not the nature of those mental states themselves. Just as van Fraassen's initial considerations underdetermined whether or not we were dealing with two distinct attitudes, or one attitude with distinct motivations, so too do Williams' considerations underdetermine whether we are dealing with two distinct attitudes, or one attitude about which an agent has distinct (and thereby erroneous) opinions. The rhetorical case for the constructive empiricist's distinction may be mounting, but we have yet to provide a conclusive argument.

Nevertheless, the considerations we have assembled so far – subtleties in the appropriate characterisation of a mental state, the basic distinction between pragmatic and credential considerations and a focus upon the sorts of understanding that an agent may have about his own mental states – suggest a final line of argument, one that involves a closer examination of the individuation of intentional activity. The most important idea here – due largely to Anscombe (1957) – is that what distinguishes intentional activity (forming or holding a particular belief, hoping or desiring for a particular state of affairs) from non-intentional activity (dropping the coffee pot, falling down the stairs) is that the former allows of a certain type of *evaluative question*. To take a simple example, it is always appropriate (within the contexts of the appropriate social norms) to ask on what grounds someone believes what they believe, to what ends someone desires what they desire, against what outcome someone fears what they fear and so on and so forth; the implication of course being for the interrogated agent to offer us a defence of his particular attitudes, that his opinions are rational. By contrast, it is rarely if ever appropriate to ask the same sort of question about non-intentional activity, such as the mishandled coffee pot or our unfortunate tumble; the sort of questions we would ask in these situations are purely factual questions about the circumstances leading up to the event, with no evaluative connotations. One asks for *reasons* for intentional activity, whereas one asks for *causes* for non-intentional activity. The two are often confused, and sometimes deliberately obfuscated: take Hilary Putnam's example of the professor who, when asked why he was found naked in the girl's dormitory at midnight, explained that he had been naked in the girl's dormitory just a couple of seconds before midnight, and given the laws of nature and their constraints on how fast a physical object can travel, could hardly have been expected to have been anywhere else at the time he was found!

Moreover, it also seems reasonable to suppose that one can individuate intentional activity in terms of the *specific* evaluative questions that it allows, and the appropriate range of answers that can be given in response. Again to take a very simple example: the belief that p naturally raises the sort of evaluative question that pertains to the truth of p, the grounds for coming to regard p as true and with whether or not the truth of p coheres with the rest of an agent's beliefs; whereas by contrast, the desire that p naturally raises the sort of evaluative question that pertains to the utility of p, with whether or not p really is in the agent's best interests and to what ends the realisation of p would serve. We can also see that, at least in the simple case above, these sets of questions will be mutually exclusive: it makes no sense to ask whether a particular desire is true; and while we may legitimately wonder as disinterested spectators whether or not a particular belief is in an agent's best interests, it similarly makes no sense to ask of the agent in question to what ends such a belief serves.

We need now to put there various considerations together. We have conceded to Horwich that the appropriate manner in which to investigate the alleged distinction between acceptance and belief is to attend to the predominantly functional characterisation of these two attitudes. We have noted however that Horwich's own understanding of the functional role of belief was too crude, and we have now seen another important dimension to this issue: we must attend not only to the causes and effects of the mental states in questions, but also how they relate to our larger interpretative projects. So not only does Horwich attribute *too much* to our notion of belief in supposing that it is either necessary or sufficient for our various utterances, and in supposing that one can give an adequate functional characterisation of belief without contrast to our more pragmatically motivated attitudes, he also attributes *too little* to our notion of belief by failing to attend to the role our mental states play in the understanding of others: part of what it means to have a belief, for example, as opposed to a desire or any other species of mental state, is that it prompts questions in others about the credibility of what we believe. With this in mind, we can offer a second line of argument for what was at the heart of Williams' account. For as Hieronymi (2006) points out, in order to believe at will, one must acquire the belief in question for predominantly *practical* reasons as opposed to the usual evidentiary considerations – for if practical reasons did not enter into the account, it is difficult to see just how such a belief could be considered voluntary, rather than just an involuntary response to the available evidence. So already we can see how the notion of a voluntary belief is in tension with our initial distinction between the pragmatic and the credential that we have found so helpful both in the functional characterisation of belief and in making sense of various aspects of our mental lives. But moreover, if one is trying to acquire a belief for practical reasons, it no longer makes sense to evaluate that activity in terms of the truth of the belief: the interesting and relevant questions will be more analogous to the sorts of evaluative questions raised by desires, that is, whether or not acquiring such a belief at will really does serve the agent's best interests. The point then is that if there was such a thing as a voluntary belief it would raise significantly different evaluative questions from what we would ordinarily classify as a belief; and since the evaluative questions raised by an intentional act are an important feature of how we individuate intentional acts – indeed, are constitutive of their identity as an intentional act – a 'voluntary belief' cannot in fact be the same thing as a belief. Williams is therefore right that there is a conceptual impossibility in the notion of believing at will; but the impossibility lies not so much in what an agent can rightly regard of himself as having done (which on Horwich's behalf we noted was an unreliable indicator), but in that such a notion contradicts the basic distinction between pragmatic and credential considerations that an interpreter must apply in understanding the agent. Of course, one may be able to *indirectly* bring about a deliberate change in one's beliefs as a result of other intentional activity, by choosing to visit a hypnotist for example; but in virtue of its role in our evaluative and interpretative practices, a voluntary belief is simply a misnomer.

My claim then is that belief, in contrast to acceptance, is essentially involuntary. One can choose what one accepts, since this is fundamentally a pragmatic attitude, but one cannot choose what one believes, since this is primarily concerned with how things strike you. My extremely expensive solicitor can choose to *accept* that I am innocent of the daring train robbery of which I am accused, since otherwise he would be unable to defend me to the best of his abilities; nevertheless, his mounting suspicion over my mysterious ability to suddenly afford his exorbitant services may make it impossible for him to *believe* that I am innocent. The argument began with whether or not an agent could regard himself as having formed a belief at will; and while this underdetermined the case at hand, a closer examination of the individuation of intentional activity – in particular, the role such activity plays in our interpretative practices – mitigated William's contention that no sense could be afforded to the notion of a voluntary belief. Crucially then, if the pragmatic aspect of acceptance entails that it is a voluntary attitude, whereas the credential aspect of belief entails that it is involuntary, Horwich cannot dismiss such an argument in the same manner as before. To argue that the proposed distinction between acceptance and belief merely illustrates two different aspects of the same attitude would be to argue that one attitude is both voluntary and involuntary, which is clearly inconsistent. A little elaboration upon van Fraassen's initial characterisation therefore yields a rather powerful response to the constructive empiricist's critics.

4.2.3 Contradiction and science

My defence of the constructive empiricist's distinction between acceptance and belief has thus far consisted of the following two lines of argument: firstly, that the case against this distinction (insofar as the constructive empiricist's critics have even bothered to make such a case) is deeply flawed, resting as it does upon both an extremely specific philosophical theory of mind and a desperately weak understanding of the nature of belief; and secondly, that van Fraassen's initial contention that the distinction between acceptance and belief is a distinction between a primarily pragmatic attitude and a primarily credential attitude that one could hold towards a proposition or theory – further elaborated in terms of whether or not one could choose to accept or believe the proposition or theory in question – not only satisfied Horwich's functionalist presuppositions, but is independently plausible in terms of making sense of various aspects of our intentional and interpretative activities. The final line of argument I wish to present considers the necessity of something like the distinction between acceptance and belief for the philosophy of science in general, quite independently of the issues surrounding our present interests in the articulation and defence of constructive empiricism.

The first point to note then, as Lipton (2007a: 119) has argued, is that there are many other positions within the familiar scientific realism debate which are also committed to what appears to be the constructive empiricist's distinction between acceptance and belief.[8] Take, for example, the epistemological version of structural realism, most famously associated with Worrall (1989). This is one of a range of views regarding the aim of science that possesses a superficial similarity to constructive empiricism insofar as it also advocates an objective less than full belief – in this case, that the aim of science is truth with respect to the *structural* elements of an accepted scientific theory, these being primarily the relationships that are said to hold between whatever it is that our scientific theories are about (as opposed to those existential claims that purport to tell us *what it is* that these relationships hold between), and usually expressed in the way of a mathematical equation. Yet in contrast to constructive empiricism, this more parsimonious assessment of the aim of science is not motivated by a broader conception of the appropriate epistemology for the philosophy of science, or as part of a larger project concerning the notion of rationality in general, but simply as a way to balance the two competing intuitions concerning the success of our scientific practices with which we began this book. So, on the one hand, the unprecedented predictive and manipulative success of our contemporary scientific theories gives us good reasons to suppose that they are at least approximately true; on the other hand, the eventual demise of all those scientific theories considered to be approximately true in the past gives us good reasons to expect our current theories to face a similar fate. The structural realist's response – and so *mutatis mutandis* for the entity realist, the semi-realist and all of the rest – is to stake out a middle ground capable of respecting both intuitions: the historical track record shows that we cannot consider our scientific theories to be true *simpliciter*, but their current success shows that they must be true *in some respect*; the central task for the structural realist is then to show that there is some plausibility in attributing the success of our scientific theories to all and only those claims that they make regarding structure.

This however cannot be the whole story; for even if the structural realist is right that the aim of science is merely accuracy at the level of structure, something must still be said about the status of the non-structural claims of our scientific theories. As Psillos (1999: 153–155) argues, the structural realist's view cannot be that we should *only* believe what our scientific theories say about the mathematical relationships that govern the physical world, and thus that we should *reject* as superfluous those claims of our scientific theories that attempt to describe the constituents of that physical world, since the structural components of our scientific theories alone cannot satisfy the structural realist's conflicting desiderata. That our scientific theories are true with respect to the structure of physical reality cannot accommodate the intuition that the predictive and manipulative success of our scientific theories give us reasons to suppose that they are approximately true, since the structural claims of a theory alone will not generate *any* predictions (or facilitate *any* manipulations) *at all*, let alone any successful ones – an uninterpreted mathematical equation tells us nothing about the physical world in the absence of some non-structural claims regarding the content of its parameters, just as Newton's equations will tell us nothing about the motion of the heavens in the absence of those non-structural claims regarding (among other things) *how many planets there are*, and more importantly, *what sort of thing a planet is* (in terms, for example, of whether or not the mass of a planet remains roughly constant, whether or not it maintains its physical cohesion and so on and so forth). In short then, if the structural realist's position is meant to give us an account of how we can indeed suppose our contemporary scientific theories to be partly (and progressively) true despite their radically discontinuous historical track record, then the non-structural claims of our scientific theories cannot be *dismissed* since they play an ineliminable role in the predictions and manipulations upon which this optimistic assessment is based; yet if the structural realist's position is also meant to give us an account of how we can indeed accommodate the radically discontinuous historical track record of our scientific theories without thereby losing all faith in our contemporary scientific success, then the non-structural claims of our scientific theories cannot be *believed* since it is in terms of the continuous overhauling of the purely descriptive apparatus of our scientific theories that this discontinuity is said to lie. So the structural realist cannot believe the non-structural claims of his successful scientific theories, but he must undertake a pragmatic commitment to them, make predictions and offer explanations on their basis and pursue a research programme within their framework, and all of the other elements that characterise the constructive empiricist's notion of acceptance. And as for the structural realist, so too for all the other positions within the philosophy of science that attempt to satisfy both the no-miracles intuition and the pessimistic meta-induction by way of restricting those aspects of our contemporary scientific theories that are believed to be true.

The constructive empiricist is therefore far from alone in requiring a principled distinction between the attitude he takes towards the claims of a scientific theory on the basis of their credibility and the attitude he takes towards the claims of a scientific theory on the basis of their utility; the distinction between acceptance and belief is a ubiquitous element of the contemporary scientific realism debate, quite independently of the constructive empiricist's particular interests in either the nature of empiricism or the nature of our standards of rationality. Moreover, there is an important sense in which something like the constructive empiricist's distinction between acceptance and belief should be a component of any satisfactory account of our contemporary scientific practice, even independently of the specifics of the scientific realism debate. What I want to argue is that in order to make sense of contemporary scientific practice, we must make sense of both the ubiquity and the importance of *contradictions* within our contemporary scientific practice; the argument being that since many of the contradictions that we encounter are often deliberate, and in many cases quite beneficial to our scientific pursuits, they cannot be dismissed as the imperfect realisation of a scientific ideal – they must be afforded a positive and constructive account. Yet since *belief* cannot be the appropriate response to a contradiction, no matter how productive, the familiar distinction between pragmatic and evidentiary considerations – between acceptance and belief – is again to be seen as a fundamental component of the philosophy of science, rather than as a mere constructive empiricist idiosyncrasy.

Some of the examples I have in mind have already appeared in the discussion of the essentially pragmatic aspect of acceptance, such as the utilisation of a manifestly false idealisation in order to facilitate certain predictions. There my emphasis was on the idea that since such an idealisation was known to be false, it could not be a candidate for belief. But another way to approach the case is to note that our grounds for considering frictionless planes and perfectly elastic collisions *as* manifestly false idealisations are that they contradict the other assumptions we hold regarding the micro-properties of solids – we have to accept such idealisations, not just because we regard them as false, but also because it is logically incoherent to add them to our existing set of beliefs. So even for those with an extremely optimistic assessment of the aim (not to mention the results) of scientific practice, belief is not always an option: one cannot both believe in the possibility of a frictionless plane and not believe in the possibility of a frictionless plane, no matter what one's opinion regarding the success of science; yet since one must employ both assumptions in the appropriate circumstances, one must be able to countenance an attitude of pragmatic commitment that is not an instance of belief. And as for the case of idealisation, so too for the broader issue of inter- and intra-theoretical comparison. As Lipton (2007a: 124–125) notes, we regularly employ different scientific theories in different domains that make fundamentally incompatible claims about the nature of reality; and we regularly employ different models of the *same* scientific theory in different domains that make fundamentally incompatible claims about the nature of reality. The former case is (arguably) illustrated by the continuing difficulties of finding a consistent reconciliation between quantum mechanics and general relativity, the physics of the very large and the physics of the very small; the latter case clearly illustrated in the case of, say, those branches of dynamics where fluids are modelled as either discrete or continuous media as the situation requires. And even though such conflicting assumptions may rarely come into direct contact, it is surely deeply unsatisfactory to attribute to practising scientists the sort of pronounced schizophrenia that would allow them to *believe* all of the above.

Any satisfactory account of scientific practice must therefore make room for the distinction between what one believes to be true and what one endorses for largely pragmatic reasons – even the full-blooded scientific realist must allow for the constructive empiricist's distinction between acceptance and belief (although of course, he may also be committed to the idea that such a distinction will drop out once we achieve the ideal limit to scientific inquiry). The case is made with particular clarity when we come to consider two of the most important themes in contemporary philosophy of science: the relationship between prediction and data; and the derivation and explanation of new phenomena.

The first of these concerns the importance of *anomalies*, data that contradicts the predictions of an accepted scientific theory. This relates to issues concerning the methodology of science, and many philosophers have stressed the importance of putting our accepted scientific theories into situations that might produce anomalous results (the idea of the 'crucial test') – and of what one should do when confronted with such anomalous results – for understanding the mechanism whereby theories are generated, accepted and rejected. Karl Popper (1959) of course made the confrontation of theory and data, and the rejection thereby of any theory falsified by anomalous results, the cornerstone of his scientific methodology; and to take the other extreme, even Thomas Kuhn (1962) believed that it was the gradual accumulation of anomalous results that eventually precipitated a scientific revolution. Yet neither account is complete without some story as to what happens in the intervening period between the accumulation of an anomalous result and the acceptance of an alternative theory. This is particularly clear in the case of Kuhn, since the deeply institutionalised nature of the dominant scientific paradigm will often take a wealth of anomalous data to dislodge; and Popper himself repeatedly stressed that it only made sense to abandon a falsified scientific theory if one had a superior alternative to take its place. Whatever one's mechanism for rejecting one scientific theory in favour of another, there will sometimes be cases when one must retain the old theory despite the existence of contradictory data – again, cases where one must retain a pragmatic commitment to the theory even though one can no longer believe it (Lipton, 2007a: 127).

The ubiquity of contradictions however is one thing; their importance is quite another. For while I have argued that to employ an idealisation in order to facilitate an inference, or to pick and choose from a range of mutually incompatible scientific theories across various different domains of application or even to take a sober assessment of one's options in the face of anomalous data require us to draw a principled distinction between that which we find useful and that which we find credible (or else face an inconsistent set of beliefs), one may well feel that these examples are all based upon the contingent limitations of scientific practice: that Popper must sometimes accept a falsified scientific theory through lack of available options or that Kuhn must sometimes let institutional inertia cloud our appreciation of a decisive result just give us reason to reject their specific accounts of scientific methodology; that our best scientific theories of the very small conflict with our best scientific theories of the very large merely shows that we have yet to examine the issues from the right perspective; and the need for idealisations tells us more about our own feeble intellectual capacities than it does about the nature of science. These examples are all due to a kind of human error, and are to be overcome rather than accommodated as an essential feature of scientific inquiry.

Yet there are some cases where the endorsement of contradictory propositions or theories has played an important role in what we would surely wish to consider as paradigmatically legitimate scientific practice. The conceptual development of quantum mechanics at the beginning of the twentieth century – undoubtedly one of the great 'success stories' of contemporary scientific inquiry – provides a range of detailed examples: I take for illustration the derivation of Planck's Radiation Law, and the excellent discussion by Smith (1988), not least because it demonstrates the difficulties in trying to explain away what appear to be contradictory beliefs. The standard histories recount two stages in the derivation of the rate of black-body radiation: the original attempt by Planck himself (1900a; more detailed elaborations appearing in his (1900b) and (1901)); and the allegedly more satisfactory derivation later provided by Einstein (1916, 1917). In both cases, inconsistent assumptions are made that treat the system in question as exhibiting both classical and quantum mechanical properties. In Planck's derivation, one makes the classical electrodynamic assumption that the radiation of energy in the system is *continuous* in order to derive an expression for its average distribution; and one makes the quantum mechanical assumption that the radiation of energy in the system is *discrete* in order to derive an expression for its dependency upon frequency and temperature: in short, one must assume that the resonators in the system are restricted to discrete energy exchanges with the electromagnetic field and also that they exchange energy continuously with the electrodynamic field (Smith, 1988: 434–436; see also Jammer, 1966: 383–385). And similarly with Einstein: one also makes the classical electrodynamic assumption that excited atoms radiate and absorb energy *continuously* in order to justify a particular claim about the overall distribution of energy in the system (in this case, that atomic systems exhibit both spontaneous and stimulated emission of radiation), and the quantum mechanical assumption that such radiation and absorption are *discrete* in order to derive the quantitative relationship sought (Smith, 1988: 436–437; for more on the intellectual development of Einstein's account, see Kuhn, 1978: 170–187).

These incidents appear to be a clear example of how the deliberate employment of inconsistent premises has played a crucial role in scientific inquiry. But the fact that there were *two* stages in the derivation of the Planck Radiation Law raises an important challenge – for as the standard histories are quick to point out, it was the fact that Planck's original derivation was generally considered a failure that provided the principal motivation for Einstein's alternative (and subsequently considered more satisfactory) approach. Moreover, there is an important difference between Planck's and Einstein's derivation of the law – for while they both make classical assumptions in order to justify a claim concerning the overall distribution of energy in a system, the distribution upon which Einstein based his derivation was one that enjoyed substantial experimental evidence. Putting these two considerations together, argues Smith (1988: 437–439), undermines any contention regarding the utility of contradictions. Planck's derivation was straightforwardly inconsistent – it bases the black-body radiation law upon two exclusively theoretical assumptions that attribute incompatible properties to the same system – and was rejected as such by the scientific community. By contrast, Einstein's derivation made only an *incidental* appeal to classical electrodynamic theory: the overall distribution of energy was an empirical fact, and while this could indeed by explained by appeal to certain classical assumptions, this explanation was not necessary to the derivation as a whole. Another way to put the point is to note that while the inconsistency in Planck's derivation looked irresolvable (since the incompatible assumptions enjoyed no further justification than the incompatible theories from which they were drawn), the inconsistency in Einstein's derivation looked to be merely temporary – since the distribution of energy to which he appealed had been experimentally verified, there was good reason to suppose that it would be explained by any *successor* to the current electrodynamic theory, which may or may not be inconsistent with the quantum mechanical assumptions involved. As Smith puts it, 'Einstein uses hypotheses that he has reason to believe must be part of any inconsistency resolving electrodynamics that would replace the inconsistent set of claims employed in early quantum theory' (1988: 437).

This then is Smith's general model for thinking about, and thereby resolving, any apparent appeal to the utility of contradictions: that in all those cases where scientists appear to endorse an inconsistent set of theories, they in fact only endorse a mutually consistent subset of those theories – the members of which are justified by either their theoretical basis *or* independent experimental evidence. To put the idea the other way around, any apparently inconsistent premises required for a particular prediction or explanation are either justified by a mutually consistent fragment of the broader theoretical background or are reasonably *expected* to be justified by any adequate successor to the broader theoretical background. Scientists do not reason from *inconsistent* theories, they reason from *incomplete* theories; and so the case for acceptance rather than belief is duly undermined.

Yet there is clearly something deficient in Smith's reconstruction. Perhaps the simplest way to express the problem is to note that the reasonable expectation that one's premises will eventually enjoy their theoretical justification from some as yet unthought-of piece of scientific reasoning still falls a long way short of the *belief* in those premises: Einstein may well have had greater justification in this expectation than Planck, but the fact still remains that he was not in any better a position to believe both the classical and the quantum mechanical assumptions upon which his derivation was based since he was not in any better a position to know anything at all about these inconsistency resolving successor theories upon which these assumptions will eventually be justified. The situation is of course made complicated by the existence of independent experimental evidence for the overall distribution of energy to which Einstein appeals, but we must still be cautious in endorsing Smith's diagnosis. Such independent evidence can play one of two roles: it can increase our subjective probability that the theoretical fragment that predicts such phenomena will be retained in our successor theories; or it can be so overwhelming that we are prepared to use the existence of such phenomena in our scientific reasoning, irrespective of whether or not we can provide it with a satisfactory theoretical underpinning. It is important however not to conflate these two roles: for while the latter case can quite straightforwardly give us reason to believe that, say, atomic systems exhibit the sort of distribution of energy that Einstein assumes, the former can only ever add to our reasonable expectations which will by their nature always fall short of belief. Yet Smith's diagnosis trades upon just such an equivocation – he argues that the independent experimental evidence available to Einstein was sufficiently conclusive as to give him reason to believe the classical assumptions that drive his derivation, yet apparently was not so conclusive that he could abandon the electrodynamic underpinnings altogether. And this is the crucial question: if the experimental evidence available to Einstein was what distinguished his successful derivation from Planck's failed attempt – in providing him with independent reasons to believe what Planck could only infer from a set of inconsistent scientific theories – then why did Einstein continue to use classical electrodynamics in his reasoning?

What distinguishes Einstein's derivation from Planck's is a matter of degree, not of kind. The very fact that Einstein continues to appeal to classical electrodynamics demonstrates that his independent experimental evidence was not sufficient to cut the Gordian knot and justify an independent belief in the types of radioactive emissions present in an atomic system. The independent evidence simply gives Einstein better justification for his reasonable expectation than that enjoyed by Planck. But as I already noted, reasonable expectation is not belief – it is a pragmatic motivation to treat certain assumptions as if they were true; it is an instance of acceptance, rather than belief. And as for the Planck/Einstein case, so too in general: one cannot explain away the apparently successful utilisation of a contradictory set of propositions or theories as an instance of the successful utilisation of a consistent subset of those propositions or theories, since any justification offered for such a diagnosis would be necessarily self-defeating. The simple fact of the matter is that in order to individuate the members of this consistent subset, one must appeal to some external factors – such as Einstein's experimental evidence – to distinguish which theoretical fragments we have reason to believe from those we do not. Yet if such external factors are indeed sufficient to justify belief in these mutually consistent fragments of our larger scientific theories, they would have been sufficient to justify belief in these mutually consistent fragments quite independently of their wider theoretical context; but if that is the case, our initially inconsistent situation would never have arisen, since we would never have bothered invoking such a problematic theoretical background if more straightforward justifications were available!

Any satisfactory philosophical account of our contemporary scientific practice must accommodate both the ubiquity and the utility of contradictions – in the employment of manifestly false idealisations; in the endorsement of mutually incompatible theories and models; in the assessment, weighing and bracketing of anomalous data; and in the derivation of some of our most important results. All of these require a more nuanced epistemological picture than that of belief and rejection; they all require a distinction between those propositions and theories that we believe because we find them credible and those propositions and theories that we accept because we find them useful. The constructive empiricist's distinction between acceptance and belief is thus a fundamental component of any satisfactory philosophy of science, irrespective of our opinions regarding constructive empiricism in general.

Moreover, just as the distinction between acceptance and belief plays an important role in understanding various issues within our scientific practices, so too does it lend itself to our understanding of various issues relating our scientific practices to other aspects of our intellectual lives. One example may be the conflict between what Wilfred Sellars called our 'manifest' and 'scientific' images of the world. Our everyday conception of the world – of the medium-sized dry goods we all know and love – seems radically at odds with the more theoretical description furnished by our contemporary scientific theories; consider Arthur Eddington's famous illustration:

> I ... have drawn up my chairs to my two tables. Two tables! Yes; there are duplicates of every object about me .... One of them has been familiar to me from earliest years .... It has extension; it is comparatively permanent; it is coloured; about all it is substantial .... Table number two is my scientific table. It does not belong to the world previously mentioned .... My scientific table is mostly emptiness. Sparsely scattered in that emptiness are numerous electric charges rushing about with great speed; but their combined bulk amounts to less than a billionth of the bulk of the table itself. Notwithstanding its strange construction it turns out to be an entirely efficient table. It supports my writing paper as satisfactory as table number one; for when I lay the paper on it the little electric particles with their headlong speed keep hitting the underside, so that the paper is maintained in shuttlecock fashion at a nearly steady level.
>
> (Eddington, 1928: xi–xii)

We cannot believe the table to be both substantial and mostly emptiness; that would be to believe a contradiction. Of course, we could always reject one of the conflicting views, or at least, explain one in terms of the other. The manifest image is strictly speaking false; it is merely the inevitable manner in which creatures such as ourselves come to *apprehend* the scientific image. But for those of us like Eddington, for whom the manifest image is just too psychologically well-entrenched, this may just not be an option – he writes that even though 'modern physics has by delicate test and remorseless logic assured me that my second scientific table is the only one which is really there', it remains the case that 'modern physics will never succeed in exorcising that first table' (Eddington, 1928: xiv). Acceptance may then be the proper response to resolving the tension between our scientific and our common-sense views of the world; and if so, it may also provide a constructive response to resolving the tensions between other conflicting views of the world, such as the religious and the secular (see Lipton, 2007a: 121–123; 2007b).

My argument in this section has been that any satisfactory philosophy of science must countenance a distinction between acceptance and belief, since any such account must accommodate the sort of substantial epistemological commitment that one takes towards contradictory propositions or theories *despite* their (intrinsic or mutual) incredibility. One cannot believe everything; but one can accept the theory and the anomalous data, the incompatible models and the inconsistent premises in one's derivations. But this is not to say that there are no constraints upon what one accepts. An interesting version of this difficulty has been raised by Lipton (2007a: 129), who notes that in a standard logical system, contradictions entail everything. If the constructive empiricist's notion of acceptance therefore allows contradictions, it would seem to follow that if one accepts an inconsistent theory then one accepts absolutely anything – for absolutely anything will be a logical consequence of what one accepts. But this is clearly not the case: even though I accept both the possibility of a frictionless plane (in order to simplify my calculations) and the impossibility of a frictionless plane (on the basis of my broader micro-physical views), I do not thereby accept any violations of the laws of conservation; even though I continue to accept both the experimental results (on the basis of the broader utility of my instruments) and the consequently falsified scientific theory (in the absence of a plausible successor theory), I do not thereby come to accept Medieval Cosmology; even though Planck and Einstein both accepted inconsistent propositions concerning the emission of radiation, they did not thereby and as a result come to accept absolutely every proposition whatsoever.

When one believes a proposition, one believes it to be true; it follows then that one has a basic rational obligation to believe the logical consequences of what one believes – for if one believes p to be true, and if p entails q, then only contingent human weakness should prevent one from believing q to be true as well (or, in the face of the gross implausibility of q, revising one's original belief in p). But no such obligation exists in the case of acceptance, which is motivated by pragmatic considerations: the truth of p may entail the truth of q, but the usefulness of treating p as if it were true has no relevance whatsoever to the usefulness of treating q as if it were true. We can capture this idea more formally by noting that while belief obeys a classical logic (in which contradictions do entail everything), acceptance obeys a *paraconsistent* logic (where a logic is said to be paraconsistent just in case it is not the case that contradictions entail everything). To offer an extremely quick gloss, the basic idea is that in a classical system, truth and falsehood are understood as being the values of a *function* that takes propositions as arguments. The function is defined to be one-one, so that a proposition cannot be both true and false. In a paraconsistent system, by contrast, propositions are understood to be *related* to the values of truth and falsehood in such a way that a proposition can be true, false, both or neither. Logical entailment is defined in the usual way, that of preserving truth. The upshot of all of this is that it is no longer possible to derive everything from a contradiction. In a classical system, an inconsistent set of premises of the form (p & ¬p) is necessarily false; thus in any inference of the form (p & ¬p) therefore q, truth is always (vacuously) preserved and thus the inference valid. But in a paraconsistent system, (p & ¬p) may be partly true (since either conjunct may be related to the value True); hence there is no necessary guarantee that an inference of the form (p & ¬p) therefore q will be truth-preserving.

It is beyond the scope of this book to discuss the issue of paraconsistent logic in detail, although one brief comment is in order.[9] This is simply to note that by endorsing a paraconsistent logic for acceptance, the constructive empiricist (and on my account, anyone who offers a satisfactory philosophy of science) is not thereby committed to dialetheism, the rather contentious view that some contradictions are true. This might be the case if the constructive empiricist was endorsing a paraconsistent logic for *belief*; however, the proposal here is simply to use a paraconsistent framework for systematising the logical relationships that exist between the propositions that one chooses to accept – that is, what one chooses to use for a variety of pragmatic reasons. All of this brings us back to our initial discussion of Horwich. Acceptance and belief differ in that one is voluntary and the other involuntary; and once this is pursued, it even transpires that the two attitudes are governed by different logics as well – there are some inferences that are valid for belief which are not valid in the case of acceptance. Their functional divergence is complete.

4.3 Empiricism and agnosticism

4.3.1 The acceptance solution

In the first half of this chapter I have argued that the constructive empiricist's second, and in many ways less prominent, distinction between acceptance and belief is actually quite robust with respect to those objections that have been raised against it. Although I have not developed the two attitudes in great detail, I have suggested that the underlying discontent with the constructive empiricist's notion of acceptance is based upon a rather simplistic (and largely functionalist) conception of the individuation of an intentional state – not to mention a rather simplistic (and again, largely functionalist) characterisation of belief itself – and therefore has yet to be developed into a compelling line of argument. Moreover, I have sketched a variety of dimensions in which the two attitudes can be more clearly distinguished: that one can *decide* what to accept, but not what to believe (since the former attitude is an essentially pragmatic and therefore inherently voluntary attitude, whereas the latter attitude is an essentially credential and thus necessarily involuntary attitude); that one can ask about the *value* of what one accepts, but not what one believes (whereas by contrast it is appropriate to ask about the *likelihood* of what one believes, but not what one accepts); and that one can accept an *inconsistent* set of propositions, although clearly one cannot believe them. Finally, I have also noted that something like the constructive empiricist's distinction between acceptance and belief is implicit within a range of other philosophical accounts of scientific practice, thus further enhancing its credibility.

It is time then to see if this unfairly maligned distinction can be put to more significant use in the articulation and defence of constructive empiricism. In particular, I want to suggest that the problem of internal coherence discussed at length in Chapter 3 – that in order to draw a distinction between the observable and the unobservable (or to provide a satisfactory account of nomological possibility and the existence of abstract mathematical objects) the constructive empiricist must violate the very position he is attempting to establish – can be resolved through liberal application of the distinction between acceptance and belief; specifically, that since the constructive empiricist need only *accept* those potentially problematic propositions crucial to drawing his various distinctions, it makes no difference that he cannot thereby *believe* them.

Let us begin then with Musgrave's original objection. In essence, the claim is that in order to draw his distinction between the observable and the unobservable consequences of our accepted scientific theories, the constructive empiricist must believe what our scientific theories say about the observability or otherwise of the various entities and processes that populate those scientific theories. Yet to believe what these scientific theories say about the *unobservability* of a particular phenomenon is to violate the constructive empiricist's central contention about the aim of science being mere empirical adequacy. In other words, in order to defend the position that in making sense of our contemporary scientific practice we need not believe our accepted scientific theories to be more than empirically adequate (i.e., we only need to believe our accepted scientific theories to the extent that they concern observable phenomena), the constructive empiricist is in fact committed to the belief that they are in fact more than empirically adequate (since in order to draw this distinction, he must also believe some of what these theories say concerning unobservable phenomena), rendering the position untenable. In response, Muller and van Fraassen have argued that to believe that a particular entity or process is unobservable does not force the constructive empiricist to believe more than what his theories say about the observable phenomena, since this belief is in fact reducible to perfectly legitimate beliefs about observables and the (exhaustive) belief that his theory of observability is empirically adequate in this respect. Yet as they concede, such a strategy can only get the constructive empiricist so far, since to believe that one's theory of observability is empirically adequate is only to believe that it correctly identifies all of the *actual* observable phenomena. But the belief that, say, electrons are unobservable must presumably go beyond this: it is to believe a certain modal claims about the unobservability of all possible electrons, not just the ones that might actually exist. And it is at this point that the appeal to epistemic voluntarism becomes paramount, since the only way for the constructive empiricist to make up this doxastic deficit is through an otherwise completely arbitrary amendment to his epistemic policy.

Implicit in Muller and van Fraassen's response however is an important concession to Musgrave: that in order to draw a satisfactory distinction between the observable and the unobservable, the constructive empiricist must *believe* those various propositions that concern the unobservability of certain phenomena. Yet with our substantive distinction between acceptance and belief now in hand, this is an assumption that we can begin to challenge. For if it transpires that the constructive empiricist does not *need* to believe such propositions, then it is of no consequence to him if he *cannot* believe them. The constructive empiricist would then be able to agree wholeheartedly with Musgrave's contention, yet continue to draw his distinction between observable and unobservable phenomena in a straightforward and non-question-begging way.

The first point to note then in defence of the idea that the constructive empiricist need only *accept* those propositions pertaining to the unobservability of unobservable phenomena is that the constructive empiricist's theory of observability (again, call it T*) is merely a *guide* to what is observable and unobservable. What determines whether or not an entity or process is observable with respect to the physiology of a particular epistemic community are objective facts about the entity or process and the community in question; the constructive empiricist merely employs his theory of observability T* to help him *know* whether or not these independent facts obtain. For the constructive empiricist 'what is observable [is] a theory-independent question. It is a function of facts about us *qua* organisms in the world' (van Fraassen, 1980: 57).

Our problematic propositions concerning the unobservability of unobservable phenomena are therefore merely propositions that the constructive empiricist must *use* in order to ascertain an independently existing metaphysical distinction. Consequently, with our substantial distinction between acceptance and belief in mind, it is not particularly clear why he needs to *believe* these propositions in order to use them for this purpose. Compare this with the more familiar case of a claim about unobservable phenomena, that electrons have negative charge for example. For the constructive empiricist, this claim also has an objective, theory-independent truth-value. Crucially though, the constructive empiricist maintains that while he does not need to believe such a claim, the attitude that he does adopt towards propositions of this kind is sufficient for subsequently using such propositions in a variety of inferences, explanations and predictions.

Arguably then, all that the constructive empiricist really needs to do in order to draw his distinction between observable and unobservable phenomena is to *accept* those propositions concerning the unobservability of the unobservable. Acceptance after all involves a substantial epistemological commitment: it 'involves a commitment to confront future phenomena by means of the conceptual resources of [that which is accepted] ... it is exhibited in the person's assumption of the role of explainer, in his willingness to answer questions *ex cathedra*' (van

Fraassen, 1980: 12). Similarly, acceptance also involves 'a commitment to a research programme, to continue the dialogue with nature in the framework of one conceptual scheme rather than another' (1980: 4), and 'a wager that all relevant phenomena can be accounted for without giving up that theory' (1980: 88). The thought then is that just as the acceptance of a proposition concerning an unobservable phenomenon is held to be sufficient for subsequently using that proposition for a variety of inferential and explanatory purposes, so too does the acceptance of a proposition about the *unobservability* of an unobservable phenomenon also suffice for subsequently using that proposition to determine which *other* propositions about scientific entities are to be believed.

This is the essence of my argument: that since the constructive empiricist is only committed to those claims concerning the unobservability of unobservable phenomena insofar as he must endorse them to draw his crucial distinction, and since the constructive empiricist has the epistemological resources to endorse the claims of his scientific theories without thereby believing them, there is no inconsistency in his position. To belabour this basic idea further would serve merely rhetorical purposes. The real test of my thesis is in its application to the more complex variations of Musgrave's objection concerning nomological possibility and the existence of abstract mathematical objects. In the simple case concerning the unobservability of unobservable phenomena, all that is required in way of a satisfactory response is to show how the constructive empiricist can distinguish between the need to use a certain proposition and the need to believe that proposition – and this is accomplished by simply reiterating the distinction between acceptance and belief that has been developed above. In the more complex cases of nomological possibility and abstract mathematical existence, however, the constructive empiricist must also show that his basic distinction between the need to use a certain proposition and the need to believe that proposition can also satisfy a variety of additional desiderata peculiar to that domain – for example, that in only accepting certain claims of nomological possibility the constructive empiricist does not thereby commit himself to an untenable philosophy of modality. If I therefore seem to have given Musgrave's original concerns regarding the unobservability of unobservable phenomena a rather curt dismissal, it is because I take the real issue of the internal coherence of constructive empiricism to lie not in whether or not the constructive empiricist can consistently maintain those distinctions fundamental to his position (which I take to be pretty straightforward, given his distinction between acceptance and belief), but with whether or not he can maintain these distinctions as part of a positive and independently plausible philosophy of science.

But before turning to these more substantial issues of modality and mathematics, I would like to add three further considerations in defence of my basic strategy. The first is to recall from my previous discussion that there is no interesting sense in which the attitude of acceptance can be thought of as being *weaker* than the attitude of belief. One worry that my strategy may have raised is that while the constructive empiricist may be able to avoid the straightforward inconsistency of needing to believe a proposition in order to establish the position that one need not believe that proposition, it doesn't necessarily follow that the constructive empiricist can thereby draw his crucial distinctions in a fully satisfactory manner: to demonstrate that he can accept the distinction between observable and unobservable phenomena rather than believing it is one thing; but if acceptance transpires to be a hopelessly impoverished attitude, the constructive empiricist can take little comfort in offering such a response to Musgrave. It is important then to reiterate that in accepting a proposition or theory one is not thereby undertaking a weaker commitment than if one had believed that proposition or theory. The two attitudes are quite distinct, rendering such comparisons meaningless. The constructive empiricist therefore is certainly doing something *different* when he accepts those claims of his theory of observability; but there is no sense in which this can be taken as a concession to Musgrave.

The second consideration concerns the special status of the putatively problematic propositions in question. Those claims of the constructive empiricist's theory of observability that concern the unobservability of unobservable phenomena determine which other claims of his accepted scientific theories that he can believe. Such claims therefore function somewhat like framework principles for the constructive empiricist: they are claims that demarcate the domain over which the question of belief can be raised. And similarly for those claims concerning nomological possibility and the existence of abstract mathematical objects – for just as the constructive empiricist needs to know which phenomena are unobservable in order to know which claims of his scientific theories he can believe, so in general does he need to know which entities or processes he would have observed, had circumstances been different; and just as the constructive empiricist needs to know which claims of his scientific theories he can believe with respect to his contention that the aim of scientific inquiry is merely empirical adequacy, so in general does he need to know what it means to have a scientific theory in the first place. The idea then is that not only does the constructive empiricist not need to believe these putatively problematic propositions about unobservability, nomological possibility and abstract mathematical existence, but moreover that such propositions cannot even be candidates for the constructive empiricist's belief since they must be presupposed before matters of belief can even arise. Such propositions can be thought of more along the lines of *methodological principles* for the constructive empiricist; and it makes little sense to ask if one *believes*, rather than *accepts*, one's methodology.

The third and final consideration concerns the broad motivation for my strategy. Arguably, the most interesting and innovative feature of constructive empiricism as a position within the philosophy of science is that it shifts the focus of anti-realism from questions of semantics to questions of epistemology. Traditional empiricist positions have attempted to defend their characteristically parsimonious opinion with respect to the aim of science by arguing that, regardless of our optimism concerning the success of contemporary scientific inquiry, belief in unobservable entities and processes cannot be warranted since our accepted scientific theories do not even include reference to unobservable phenomena in the first place. These arguments go hand-in-hand with some strategy for reinterpreting the surface grammar of those scientific theories that do apparently make such reference. So for example, one could attempt to eliminate putative reference to unobservable phenomena by reducing any claim purporting to be about the unobservable to a complex of claims that only concern observable phenomena (e.g., that something has a specific temperature just in case, if it is placed next to a thermometer, then the mercury rises to a specific height). Another proposal is that those claims that purport to refer to unobservable entities or processes are not strictly speaking referential devices at all, but rather meaningless syntactic constructs used for the systematisation of observable predictions. All such strategies however have been deemed to have failed, on the grounds that our unobservable vocabulary is both *indispensable* to our understanding of scientific practice (making ineliminable appearance in the predictions we make, the explanations we offer and the research programmes we formulate) and has irreducibly *excess content* over any attempted reduction.

The constructive empiricist avoids these problems since he is prepared to take those claims of our accepted scientific theories concerning unobservable phenomena at face value. Nevertheless, he can continue to defend the view that the aim of science is empirical adequacy rather than truth, on the grounds that there is an *epistemological* distinction to be drawn between the observable and the unobservable consequences of our accepted scientific theories such that the constructive empiricist can believe the former but not the latter. The substantive question that has been the main focus of this book concerns how exactly this epistemological distinction is to be articulated. According to van Fraassen, the constructive empiricist can believe what his scientific theories say about observable phenomena while refusing to believe what they say about unobservable phenomena because at the end of the day our rational obligations are extremely meagre; we have however seen reasons to be sceptical of this view. Nevertheless, any alternative proposal should respect van Fraassen's basic contention – that the empiricist view regarding the aim of science must be defended upon epistemological grounds. My contention then is that since the application of the distinction between acceptance and belief is a straightforwardly epistemological strategy – one that focuses upon the variety of *attitudes* at the constructive empiricist's disposal, if not the variety of *rationalities* – this is a solution very much in the ethos of constructive empiricism. Indeed, given the importance of the shift from semantics to epistemology for constructive empiricism in general, it is remarkable that the various responses offered by van Fraassen and his collaborators to the various problems of internal coherence are also importantly *semantic*: beliefs about unobservables are reducible to beliefs about observables; counterfactual conditionals are reducible to the logical consequences of our scientific theories; existential claims concerning abstract mathematical objects are reducible to ontologically neutral claims concerning the consistency of a particular mathematical fiction. This in fact provides another broad diagnosis of why such strategies were bound to fail. However, the acceptance solution is by contrast is an independently attractive strategy for the constructive empiricist to adopt – irrespective of the weaknesses of van Fraassen's own strategies – since it is one that fully realises the epistemological focus that is the definitive feature of his philosophy of science.

4.3.2 Committed modal agnosticism

The acceptance solution thus provides a simple and well-motivated response to the problem of internal coherence raised by Musgrave: if the constructive empiricist need only accept what his scientific theories tell him about the unobservability of unobservable phenomena, then it is of no pressing concern that he cannot also believe what they tell him. The difficulty is thus resolved by the natural application of an independently plausible distinction between the different attitudes that one can hold towards the consequences of our successful scientific theories, rather than an arbitrary amendment to the constructive empiricist's epistemic policy that can only be justified within the somewhat contentious contexts of van Fraassen's epistemic voluntarism.

A similar solution also suggests itself with respect to Ladyman's concerns regarding the constructive empiricist's account of modality – not surprisingly perhaps, given my contention in Chapter 3 that the problems raised by Musgrave and Ladyman (and indeed by Rosen, to be discussed in the next section) are structurally equivalent and admit of a common diagnosis. In essence, Ladyman argues that constructive empiricism is incompatible with the deflationary metaphysics that motivates it. Specifically, he argues that in order to draw a principled distinction between observable and unobservable phenomena, the constructive empiricist must be committed to believing some of the modal implications of his accepted scientific theories: this is because there will be some phenomena that will never actually be observed, yet which we would intuitively still wish to classify as observable. The constructive empiricist is thus committed to believing certain counterfactuals, and thus believing some of the modal implications of his accepted scientific theories. The dilemma for the constructive empiricist is whether or not being an observ*able* phenomenon is an objective modal fact. For on the one hand, if there were no objective modal facts, then whether or not something counted as an observable phenomenon would depend upon which scientific theory we use to describe it; and if this were the case, the constructive empiricist's distinction between the observable and the unobservable would be too arbitrary to sustain his philosophical position. Yet on the other hand, if there are objective modal facts, then the constructive empiricist's distinction between the observable and the unobservable rests upon exactly those unobservable phenomena that he claims are unnecessary for understanding the aim of science; and if this were the case, the constructive empiricist's distinction between the observable and the unobservable would be so robust as to undermine his philosophical position.

Monton and van Fraassen's response is to attempt to justify the arbitrariness of their modal nominalism by recourse to the minimal standards of epistemic voluntarism, in much the same way – and to much the same effect – as Muller and van Fraassen attempt to justify their response to Musgrave. And in much the same way, the acceptance solution opens up a much needed third possibility. Ladyman's central premise is that in order to distinguish between observable and unobservable phenomena, the constructive empiricist is committed to believing some of the modal implications of his accepted scientific theories. It is this premise that we are now in a position to deny. In order to draw his distinction, the constructive empiricist is certainly committed to various counterfactuals. Moreover, in order for this distinction to be non-arbitrary, these counterfactuals must indeed have objective and theory-independent truth-conditions. Yet it does not follow that the constructive empiricist is committed to the beliefs that Ladyman assumes, nor that he is drawn into the dilemma Ladyman presents. In fact, there are *two* ways to resist this reasoning. For on the one hand, the constructive empiricist can argue that his beliefs towards the modal implications of his scientific theories are not *existentially committing*, and therefore do not generate Ladyman's dilemma; on the other hand (the acceptance solution, and the strategy that I will favour), he can argue that his modal commitments can be satisfied by an attitude *other than belief*.

The first response draws upon some recent work by Divers (2004; 2006), and which for reasons that will soon become obvious I shall refer to as *uncommitted* modal agnosticism. The basic idea here is that agnosticism about the existence of other possible worlds – in conjunction, that is, with a Lewisian analysis of our modal discourse – does not entail wholesale agnosticism about modality. As Divers notes (2004: 669–673), much of our modal discourse is concerned with making unrestricted, negative existential claims (that is to say, many of our modal assertions are of the form that 'there is no possible world where . . . '), and much of this can be known to be true or false *irrespective* of our belief in the existence of other possible worlds. To take a simple example, the claim that a proposition p is necessarily true can be analysed as saying that there is no possible world where p is false. It follows then that if we know p to be false at the actual world (and therefore that p is false for at least one possible world), we can know that the original claim is *false* – and all this irrespective of our belief in the existence of other possible worlds. And conversely, if we know that p is a logical truth, we can then know that p is necessarily true since it follows that there will be no world where p is false *irrespective of whether or not any other possible world exists*. For our present purposes, it is important to note that a similar story can also be told about counterfactuals. As Divers argues (2004: 671–673), on the standard Lewisian account, a counterfactual claiming that if A had been the case, then C would have been the case is to be analysed as saying that there is no (relevantly similar) possible world where A is true and C is false. For similar reasons to the ones sketched above then, it follows that a modal agnostic can know whether certain counterfactuals are true or false, regardless of the existence of other possible worlds. Again to take a simple example, if we know that the actual world is one where A is true but C is false, then we know that there is at least one relevantly similar possible world where A is true and C is false; and thus we know the counterfactual to be *false*. And conversely, if we know that our antecedent A is impossible, then we know the counterfactual to be (trivially) true, since there will be no relevantly similar worlds where A is true and C is false – since there will be no relevantly similar worlds where A is true – *irrespective of whether or not any other possible worlds exist*.

All this then suggests the following response to Ladyman. The constructive empiricist can concede that he is indeed committed to believing certain counterfactuals in order to draw his distinction between observable and unobservable phenomena, and moreover that these counterfactuals must be given objective, theory-independent truth-conditions. However, he can then argue that since the various counterfactuals in which he is interested are all in fact to be analysed as negative existential claims, he can know their truth-values *without* endorsing the existence of other possible worlds (i.e., possible worlds besides the actual world).[10] Consequently, the constructive empiricist can deny that his counterfactual truth-conditions are entirely arbitrary, since he is endorsing a full-blooded realist analysis of our modal discourse (rather than an account that ties our counterfactual truth-conditions to our preferred conventions about how we are to represent the world); and moreover, he can also deny that it is inconsistent for him to know the truth-values of these counterfactuals, since such knowledge does not in fact require the existence of philosophically troublesome worlds, and hence does not require any metaphysically robust similarity ordering of these worlds.

Nevertheless, there does appear to be a serious shortcoming with uncommitted modal agnosticism. The problem lies with the sort of modal statement that the uncommitted modal agnostic *cannot* know the truth-value of. If the constructive empiricist wishes to draw a principled distinction between observable and unobservable phenomena, he is not just committed to those counterfactuals of the form that if A had been the case, then C would have been the case: he must also be committed to *denying* various counterfactuals of the form that if A had been the case, then C *would not* have been the case. For example, in order to establish the observability of the moons of Jupiter, the constructive empiricist cannot simply affirm the counterfactual conditional that if we had travelled deep enough into outer space then we would have observed them (which on the Lewisian analysis amounts to the claim that there is no relevantly similar possible world where we travel into space and fail to observe the moons); he must also deny the contrary counterfactual conditional that if we had travelled deep enough into outer space then we would not have observed them (which, on the Lewisian analysis, amounts to the claim that there is a possible world where we travel into space and succeed in observing the moons).[11] That the first counterfactual is insufficient follows precisely from the fact that it may indeed be vacuously true: maybe we know that there is no possible world where we travel into outer space and fail to observe the moons of Jupiter; yet unless we also deny the second counterfactual – and thus assert the existence of a possible world where we *do* travel into outer space and observe the moons – we cannot ensure interplanetary observability, since we also need to know that it is possible to travel sufficiently far into outer space. If this were not the case, then the constructive empiricist's distinction would be too broad, since it would deem as observable anything whose conditions of observability lie outside of our nomological possibilities. It is only by affirming and denying contrary pairs of counterfactual conditionals that the constructive empiricist – or indeed anyone with a vested interest in their counterfactual truth-conditions – can hope to secure their appropriate modal scope.

But even if the truth-value of the first counterfactual is something that can be known without existential commitment to other possible worlds, the denial of the second counterfactual is definitely a *positive* existential claim, and hence not something that the uncommitted modal agnostic can maintain. The best that he can maintain is to argue that although he cannot deny a counterfactual conditional of the form that if A had been the case then C would not have been the case, it will often be the case that he will lack sufficient warrant to assert it – and that is hopefully enough to distinguish between those possibilities that we would wish to hold as true and those that we must hold as (vacuously) true. But to lack sufficient warrant to assert such a counterfactual is a far cry from being in a position to deny such a counterfactual, and we might again worry that such a strategy will fail to accurately specify the constructive empiricist's distinction since the range of possibilities that we lack the appropriate warrant to assert will not precisely match the range of possibilities we would wish to deny. Arguably then, since in order to draw his distinction between observable and unobservable phenomena the constructive empiricist needs to both assert and deny various counterfactual conditionals, it looks as if uncommitted modal agnosticism can only offer a partial response to Ladyman's dilemma.[12]

Uncommitted modal agnosticism appeared to offer a promising strategy for the constructive empiricist to adopt because it conceded a realist semantics for his modal discourse, yet did so in a way that undercut the existential implications that drive Ladyman's dilemma. Unfortunately, such a strategy only works for a limited domain of modal statements. It does however provide a useful contrast for the second strategy canvassed above: that the constructive empiricist concedes that he is committed to certain objectively construed counterfactual conditionals – including those counterfactuals that are existentially committing – but denies that he is thereby committed to *believing* these counterfactuals, since the modal commitments forced upon him can be satisfied by an attitude other than belief.

Consider again the sense in which the constructive empiricist is 'committed' to the modal implications of his theories: for certain phenomena, in order to determine whether or not they count as observable, he must base his judgement upon what would happen in certain (non-actual) circumstances. The constructive empiricist must therefore *use* various modal statements in order to distinguish between the observable and the unobservable. But it is not clear why the constructive empiricist needs to *believe* these statements. The situation here is exactly the same as it is with respect to Musgrave's objection, and with whether or not the constructive empiricist needs to believe those propositions concerning the unobservability of unobservable phenomena (regardless of their modal status). For just as the constructive empiricist need not believe what his scientific theories tell us about unobservable phenomena (the charge on an electron, say) in order to use those claims in his explanatory and predictive practices, he need not believe either what his scientific theories tell us about the unobservability of unobservable phenomena in order to use those claims in determining the extent of his scientific beliefs. Consequently, the constructive empiricist need not believe what his scientific theories tell us about the mere possibility of observing various phenomena in order to use those claims in determining the extent of his scientific beliefs; and by extension then, he need not believe what his scientific theories tell us about the merely possible in general. Instead, the constructive empiricist can adopt what we can think of as a *committed* agnosticism towards the modal implications of his scientific theories (in contrast to Divers' uncommitted variant), and this will be found to be sufficient for subsequently using these implications for a variety of purposes.

Essentially then, the constructive empiricist can endorse the same combination of literal (face value, realist) semantics and epistemological commitment towards other possible worlds as he does towards actual unobservable phenomena such as electrons. And similarly, he is also to refuse to assert whether or not these other possible worlds really exist, and thereby maintain an attitude of committed agnosticism towards the modal implications of his scientific theories. This gives the constructive empiricist's counterfactual conditionals objective and theory-independent truth-conditions, thus avoiding the charge of arbitrariness. Moreover, since the constructive empiricist remains agnostic about the existence of other possible worlds, he does not incur the inflationary metaphysics of modal realism, and hence avoids the collapse into structural realism. Finally, and in contrast to *uncommitted* modal agnosticism, since the constructive empiricist endorses a substantial epistemological commitment in accepting the modal implications of his scientific theories, he earns the right to *use* these implications for his various purposes without thereby committing himself to *believing* the modal implications of his scientific theories. Thus again, Ladyman's dilemma is undermined.

Ladyman identifies committed modal agnosticism in his original paper (2000: 846–847), but rejects it as a possible strategy for the constructive empiricist. Ladyman's main complaint is that committed modal agnosticism does not 'involve belief in any modal statements objectively construed', whereas the problem of modality shows that the constructive empiricist 'ought to be positively committed to there being objective relations between the actual and the possible' (2000: 849). But as I hope is beginning to become clear, this belief constraint is too severe: whether or not the constructive empiricist *believes* that these various modal truth-conditions are met is irrelevant, provided he is suitably committed to (that is to say, he *accepts*) the claim that they are. To respond directly to Ladyman then, the constructive empiricist can concede that committed modal agnosticism does not involve belief in any modal statement objectively construed, but that this is not a problem, since one can be 'positively committed' to this objectivity without thereby believing it.[13]

A second line of criticism can be found in Divers (2004: 677), who considers committed modal agnosticism, and investigates whether or not countenancing what he considers to be the weaker epistemic attitude of committed agnosticism (i.e., acceptance) can help to circumvent some of the deficits in what the modal agnostic can assert. Divers, however, rejects this approach for two important reasons. The first is simply doubt over whether or not the distinction between believing a proposition or theory and being committedly agnostic towards that proposition or theory can be plausibly maintained. This is of course the general worry with which we began this chapter and discussed in the context of Horwich's functionalism, and which I now take to have been answered. Diver's second objection, however, is one of motivation. He argues that since one can know the truth-value of various (negative existential) modal statements, there is little motivation for the wholesale modal scepticism entailed by committed modal agnosticism – such a strategy is just an epistemic concession too far. However, the primary motivation for this position is not modal scepticism, as it is methodological continuity. The proposal is for the constructive empiricist to simply extend his attitude towards actual unobservable phenomena to non-actual phenomena. Committed modal agnosticism can therefore be seen as a natural extension of constructive empiricism, and is therefore primarily motivated by its ability to provide a simple and economical response to Ladyman's dilemma. Moreover, and perhaps more importantly, committed modal agnosticism does not entail wholesale modal scepticism. The position is compatible with much of the modal knowledge that Divers mentions. After all, the constructive empiricist who adopts committed modal agnosticism still believes statements about actual observable phenomena. Consequently, any of the modal knowledge Divers mentions that is based upon actual observable phenomena will still count as legitimate knowledge for the committed modal agnostic: if it is the case that p is true in the actual world (and of course if p is about observable phenomena), then the constructive empiricist who endorses committed modal agnosticism is also in a position to know that necessarily not-p is *false*, and for exactly the same reasons that Divers gives.

Committed modal agnosticism thus provides at least as good a response to Ladyman's dilemma as uncommitted modal agnosticism. But crucially, it also provides a far more successful strategy for dealing with the sorts of modal statement that the modal agnostic *cannot* know the truth-value of. Recall that if the constructive empiricist wishes to draw a principled distinction between observable and unobservable phenomena, not only is he committed to various counterfactual conditionals, he is also committed to *denying* various counterfactual conditionals. However, since the correct analysis of these second sorts of counterfactuals involves a positive existential claim, they are not something that the modal agnostic can know the truth-value of, and thus they are not counterfactuals that the uncommitted modal agnostic can deny. For the *committed* modal agnostic, however, the falsity of the second counterfactual is just as easily endorsed as the truth of the first. For although the committed modal agnostic no more *believes* in the existence of other possible worlds than the uncommitted modal agnostic does, he does *accept* the existence of various possible worlds (those entailed by those scientific theories he believes to be empirically adequate); in particular then, he can come to accept the existence of a (relevantly similar) possible world where both A and C are true, which allows him to *both* affirm that if A had been true then C would have been true *and* deny that if A had been true then C would have been false.

4.3.3   Acceptance and anti-realism in the philosophy of modality

My contention then is that Ladyman's dilemma can be resolved in exactly the same way as Musgrave's objection can be resolved – one points out that the constructive empiricist need not *believe* those putatively problematic propositions concerning the unobservability of unobservable entities, or the counterfactual consequences of one's distinction between the observable and the unobservable, since there is another workable attitude in the offing. One then reassures the critic that there *really is* a principled distinction between acceptance and belief, and argues that since the aforementioned propositions play something like a framework constitutive role, it is in fact perfectly reasonable to only accept them. Yet while a satisfactory response to Musgrave needs to go no further than this, the situation is slightly more complicated with respect to Ladyman's dilemma. Committed modal agnosticism is a general proposal for how the constructive empiricist should approach the philosophy of modality; in addition then to any concerns one may have over the acceptance response in general, we also require the constructive empiricist to demonstrate that his position is not undermined by an implausible account of our modal discourse.

Anti-realism about modality is a relatively underdeveloped field, but among those philosophers interested in somehow 'rehabilitating' a possible worlds analysis of our modal discourse (as opposed to those who wish simply to eliminate it) the most popular options all place the weight of their anti-realism upon semantic considerations. To take the two most salient examples: the modal fictionalist argues that possible worlds discourse is anti-realistically respectable because it is really about the consequences of a useful *story* that we have about possible worlds, rather than about the possible worlds themselves; in contrast, the quasi-realist argues that our possible worlds discourse is anti-realistically respectable because it is really just an *expression* of our conceptual limitations, rather than a series of claims about metaphysically troublesome entities. And in both cases, the semantic sophistication demanded of the modal anti-realist generates serious difficulties. The argument I wish to present then is a very simple one: *if* we take a possible worlds semantics to be a desirable component of our philosophical analysis of modality, and *if* we therefore take the ability to successfully deliver such a semantics to be a fundamental desiderata of any anti-realist programme in the philosophy of modality, *then* committed modal agnosticism is not only a perfectly satisfactory position with which the constructive empiricist is to be lumbered, it is in fact the most attractive position for him to adopt quite independently of the issue of internal coherence. This does of course leave many important questions unanswered, such as the desirability of a possible worlds semantics in general. I hope however that my motivation is fairly transparent: one of the most important and interesting aspects of the constructive empiricist position within the philosophy of science, arguably, is that it relocates the realist/anti-realist debate from abstruse issues of semantics to robust considerations of epistemology – part of my project then is to investigate the prospects of a similar transition within the philosophy of modality.

To begin then with modal fictionalism: on this view, there are no possible worlds besides the actual world, and therefore all statements concerning other possible worlds are literally false (or trivially true in the case of universal generalisations). However, there remains an important distinction that can be drawn between all of these false statements: some statements about possible worlds are true *according to a certain fiction* that we can tell about possible worlds, and in this lies the distinctive elements of modal fictionalism. Rosen (1990: 330–333) draws the analogy with other, more familiar cases of statements about fiction: we can say that it is true that there is a possible world with talking donkeys in much the same way as we would say that it is true that Sherlock Holmes lives at 221b Baker Street. In more general terms, a modal statement p is true iff (the appropriate possible worlds paraphrase of) p is true *according to the appropriate possible worlds fiction* (cf. Rosen, 1990: 335).

The precise consequences of this view for the philosophy of modality, however, depend upon the degree of fictionalism that one is willing to endorse. Following Nolan (1997: 261–264), it is therefore helpful to distinguish between three varieties of modal fictionalism: broad, timid and strong. Rather schematically, the broad modal fictionalist is a fictionalist about both our talk of possible worlds *and* the modal discourse such talk is meant to illuminate: not only are there no possible worlds besides the actual world, but our talk about possibility and necessity is also – strictly speaking – false. The timid modal fictionalist is a fictionalist about our talk of possible worlds, but argues that such talk is *justified* by our modal discourse (about which he is not a fictionalist). And the strong modal fictionalist is a fictionalist about our talk of possible worlds, but argues that such talk *justifies* our modal discourse (about which he is not a fictionalist).

Of these three positions, it is only strong modal fictionalism with which we need be concerned. Strong modal fictionalism attempts to give an account of modality in terms of an anti-realist understanding of possible worlds, and thus constitutes a distinctive anti-realist position within the philosophy of modality. By contrast, broad modal fictionalism – since it is the view that both possible worlds talk and our modal discourse is false – is better classified as a form of *eliminativism* about modality; and timid modal fictionalism – since it attempts to justify possible worlds talk in terms of an independently held view about modality in general – is not really a fictionalist position in any interesting sense of the term, since all the important philosophical work will be done by whatever position the timid modal fictionalist takes towards modality in general. In what follows then, I shall refer exclusively to the strong formulation when I discuss modal fictionalism.

Since my argument in this section is for the rather specific claim that committed modal agnosticism – the *acceptance* of other possible worlds – is the anti-realist's best hope for rehabilitating a possible worlds semantics for his modal discourse, I shall not attempt a detailed evaluation of modal fictionalism.[14] Rather, I want to begin by discussing the modal fictionalist's prospects for providing a conceptual analysis of our modal discourse. According to the modal fictionalist, a claim of the form 'possibly p' is to be understood as really being of the form 'according to the fiction, there is a possible world where p'. The issue here therefore hinges on *which* specific fiction about possible worlds the modal fictionalist wishes to tell; and this of course raises the question of justifying one fiction over another. For although such a decision will presumably be constrained by both facts about the actual world and some kind of reflective equilibrium with our intuitive modal judgements, it is far from clear whether this will specify a unique possible worlds fiction. The decision then to analyse our modal discourse in terms of a particular story that we care to tell about possible worlds begins therefore to look somewhat arbitrary. Moreover, given that the existence of any particular fiction will be a contingent matter, dependant upon its formulation by some suitable inclined philosopher, modal fictionalism appears to be committed to the view that if, say, David Lewis had never existed, no positive modal statement would have been true! As Nolan (1997: 265) remarks, 'however important one thinks philosophers might be, one should not think that they are *that* important.'

More importantly, however, is the issue of the *completeness* of the possible worlds fiction endorsed. Our modal discourse consists of a potential infinity of modal statements; any possible worlds fiction that purports to give an analysis of our modal discourse must therefore have sufficient resources to analyse infinitely many modal statements. Yet clearly no *actual* fiction will have such resources: fictions are human constructions, and human construction is sadly all too limited. If the modal fictionalist is to have any hope then of providing an analysis of our modal discourse in terms of a fiction about possible worlds, he is either going to have to appeal to some *possible* fiction that one *could* formulate or going to specify some general conditions for an appropriate fiction and then appeal to the *consequences* of this initial specification as an analysis of our modal discourse. Either way of course the modal fictionalist is forced to presuppose a modal notion in order to make his fiction sufficient – either because he is appealing to a possible fiction, or because the notion of consequence it itself modal – and thus any attempt to analyse modal notions in terms of the fiction ends up looking hopelessly circular (cf. Nolan, 1997: 266–268). In short, modal fictionalism cannot provide a conceptual analysis of our modal discourse in terms of a story that we can tell about other possible worlds: for not only will any such analysis seem arbitrary and unjustified, it will also be either radically incomplete or will presuppose the very modal notions it seeks to analyse.

The prospects for a conceptual analysis look somewhat more promising for the quasi-realist however who, in contrast to the modal fictionalist, does not attempt to give an anti-realist re-interpretation of our possible worlds discourse: no anti-realist operators are introduced, no alternative locutions are provided. In fact, the quasi-realist takes pride in leaving the surface structure of our possible worlds discourse entirely untouched. The key move for the quasi-realist is rather to provide an anti-realist *justification* of such talk, some story whereby one can come to speak just like a modal realist, but without ever having gotten one's hands dirty with all that ontology. For the quasi-realist then, the issue of realism and anti-realism about a particular domain of discourse is not to be located at the level of the semantics that one uses and the assertions that one makes; rather, it is to be located in the various accounts one might give as to how one is entitled to use those semantics and to make those assertions. As Blackburn puts it, 'it is not what you end up saying, but how you get to say it, that defines your "ism" ' (1993: 7).

The basic modal quasi-realist strategy thus consists of the following two moves. The first is *projectivism*, whereby the various modal claims we make are explained as really expressing certain commitments and attitudes that we hold about how we structure our beliefs about the actual world. To claim that something is a logical necessity, for example, is to express 'our own inability to make anything of a possible way of thinking which denies it' (Blackburn, 1984: 217). There is no necessity in nature, no other possible worlds, only a projection of our own conceptual limitations which comes to take on a propositional, truth-conditional structure – and which eventually comes to be understood in terms of a fully extensional possible worlds semantics. This then is the second move, the quasi-realism proper: showing how we can start off with our commitments and expressions of attitude, and end up with the sort of semantic apparatus usually associated with modal realism. This in turn is accomplished by showing how a possible worlds discourse might become useful or indispensable to those who start off simply expressing their conceptual limitations. For example, the ability to recognise how belief varies with circumstance is certainly vital for a large portion of our cognitive lives: it underpins our appreciation of why others may disagree with us, how we ourselves might be wrong about our own opinions and how such matters could be investigated and rectified. But if an agent can come to recognise this, then arguably he has grasped what we might as well call the 'contingency' of our various beliefs; and once he has this concept, it looks as if he can begin to imitate most of the modal realist's discourse (Blackburn, 1987). Eventually, these terms and locutions – what appears to be modal realist discourse about necessity and possibility – become sufficiently ingrained and well-entrenched that the pragmatics of communication leads them to take on propositional, truth-conditional form. And then finally, this modal discourse comes to be understood as talk about other possible worlds, and for exactly the same semantic motivations offered by the modal realist. Crucially though, whereas this process leads the modal realist to endorse an ontology of possible worlds (since he took his modal discourse to be propositional from the start), the quasi-realist notes that each semantic step taken was nothing more than a pragmatically motivated refinement upon the metaphysically innocuous underlying activity of expressing our own conceptual limitations.

One immediate concern with all of this is that the quasi-realist programme looks hopelessly circular. What we are after is an anti-realist account of our possible worlds discourse, and the quasi-realist suggests that this can be sustained in terms of a (fairly sophisticated) practice of merely expressing our own limitations. Yet when we get to the level of these commitments and attitudes, we find ourselves employing those very modal notions that were supposedly under analysis: that there is no possible world where p is meant to explicate the claim that p is logically impossible, and logical impossibility is explained in terms of our *inability* to make anything of p – that is to say, in terms of what an agent can and cannot do. The worry then is that modal quasi-realism fails before it even begins, as its explanations presuppose what they are meant to explain (cf. Rosen, 1998: 389, n. 7).

However, the quasi-realist could, with some justification, offer the following response to this problem: even though our modal discourse reflects nothing more than the expression of certain attitudes, it has become so ubiquitous and well-entrenched that we now lack the capacity to describe the world without using such modal notions. Blackburn (1984: 211–212) suggests this response with a comparison with Hume: it is not an objection to (what Blackburn takes to be) Hume's causal projectivism that our brute regularities must themselves be described in causal terms (e.g., the constant conjunction of the ball *striking* the window and the glass *breaking*) since given the ubiquity of our causal discourse 'there is no way to explain our causal sayings as projections generated by something else, for there [is] no stripped [vocabulary] in which to identify the something else.' The fact that the quasi-realist's explanation for our modal discourse is itself couched in modal language could therefore be used to demonstrate not that modal quasi-realism cannot succeed, but that it has *already* succeeded.[15]

A more serious objection is that if our modal discourse – and specifically, our modal discourse as explicated in terms of the existence of other possible worlds – is essentially an expression of our own conceptual limitations, then the limits of such discourse will be constrained by the limits of what we can express. The problem here is thus analogous to the problem facing the modal fictionalist: if the existence of other possible worlds is to be grounded by the existence of a suitable fiction or set of expressions, then we need to make sure that there is a big enough fiction or large enough set of expressions to ground all of the possible worlds talk that we want. And just as no *actual* fiction will suffice, neither will the *actual* and all-too-finite expressions of our conceptual limits. Thus even though the quasi-realist may be able to finesse the possible circularity in his explanation of our modal discourse, he is still forced to presuppose a notion of modality since he is forced to ground our modal discourse in the *possible* expressions we might make about our conceptual limitations on pain of a hopelessly impoverished semantics.

Again, however, it looks as though the quasi-realist may be able to avoid this objection. If our possible worlds discourse is to be understood simply as our expression of certain attitudes, the quasi-realist can argue, there can be no sense in which the number of possible worlds that need to be grounded can outstrip the number of actual, finite expressions that we make. For unlike the modal fictionalist, who attempted to provide anti-realistically respectable substitutes for all claims about other possible worlds, the quasi-realist is in fact offering an alternative conception of what it is for us to make a claim about another possible world in the first place. From the modal fictionalist's perspective, there is an antecedently existing set of possible worlds statements that need to be accommodated; from the quasi-realist's perspective, possible worlds statement only exist insofar as we have made the corresponding expressions of attitude. The quasi-realist is therefore in a position to maintain that the range of possible worlds statements that need to be accommodated necessarily cannot outstrip the resources he has available for accommodating them; and with this proviso, it looks as if quasi-realism can provide a satisfactory conceptual analysis of our modal discourse.

Modal fictionalism cannot provide a conceptual analysis of our modal discourse, since it must presuppose such notions in order to ground everything we might want to say; modal quasi-realism may be able to provide such an analysis, but only on the basis of a substantial re-conceptualisation of what it is that needs to be analysed. By contrast, the position that I have called committed modal agnosticism can readily furnish us with what we desire.

Let me begin by briefly reiterating the committed modal agnostic position. Essentially, committed modal agnosticism stands to modal realism as constructive empiricism stands to scientific realism. That is to say, according to the committed modal agnostic, our modal discourse is to be understood as making claims about what happens at other possible worlds, and that therefore such claims are determinately true or false, depending upon the nature of these other possible worlds. However, the committed modal agnostic refuses to assert whether or not these possible worlds really exist; this is the matter over which he remains agnostic. The committed modal agnostic is therefore not ontologically committed to these entities in just the same way that the constructive empiricist is not ontologically committed to the existence of electrons. Nevertheless, the committed modal agnostic can cite numerous theoretical and pragmatic reasons for *accepting* the existence of these worlds (as opposed to *believing* in the existence of these worlds); and this, I have argued, is sufficient for being able to use our modal statements in all of the various ways we might wish.

In order to see how it is that committed modal agnostic can provide a conceptual analysis of our modal discourse, it is helpful to first recall the contrast between the various forms of modal *realism* to be found within the literature. Roughly speaking, modal realists come in two flavours. The first takes possible worlds to be non-actual, concrete entities of the same kind as the actual world; the principle exponent of this view is of course Lewis (1986). Call these the *genuine realists*. The second takes possible worlds to be actual, abstract entities of a different kind to that of the actual world: for example, as sets of propositions (Adams, 1974) or unactualised states of affairs (Plantinga, 1974). Call these the *actualist realists*. Now one of the most interesting trade-offs between genuine and actualist realism is the balance between an (allegedly) attractive ontology and the conceptual capacity to analyse our modal discourse. The genuine realist is committed to a variety of non-actual entities, and the infamous incredulous stare such entities encourage; by contrast, the actualist realist only countenances the existence of the actual world we all know and love (albeit an actual world populated with a variety of curious abstracta). On the other hand though, by limiting his ontology to the actual world in this way, the actualist realist has difficulty providing sufficient material for all the possible worlds we want. Not all sets of propositions can be identified with possible worlds, only those that are both *maximal* and *consistent* will do. But if the actualist realist is going to analyse our modal statements into claims about maximally consistent sets, he has of course presupposed the modal notions he was supposed to be analysing: a set of propositions is maximal iff it is not possible to add another proposition to the set without contradiction or repetition; and a set of propositions is consistent iff it is possible for all of its members to be true at the same time. The genuine realist by contrast, since he has a whole pluraverse of other possible worlds at his disposal, can allow a non-circular analysis of our modal discourse.[16]

This is not the place to pursue such a line of argument; what I wish to stress however is the parallel between committed modal agnosticism and the other forms of modal anti-realism discussed in this section.[17] Quite simply, the modal fictionalist failed to provide a conceptual analysis of our modal discourse because he attempted to somehow ground our modal discourse in facts about the actual world: specifically, with the stories we tell about other possible worlds. Yet since such actual bases are insufficient to ground our entire modal discourse, possible fictions were invoked, rendering any conceptual analysis hopelessly circular. Similarly, in grounding our modal discourse in the actual expressions we make of our conceptual limitations, the quasi-realist also seemed lumbered with an incomplete account that presupposes the very modal notions supposedly under analysis – although we have seen how these shortcomings may be justified on the quasi-realist's account of what our modal discourse amounts to. At least initially though, both the modal fictionalist and the modal quasi-realist face exactly the same difficulties as the actualist realist. By contrast, the committed modal agnostic is in the same boat as is the genuine realist: for him, our modal discourse is also about other non-actual, concrete worlds (albeit other non-actual, concrete worlds which he merely accepts). The committed modal agnostic can thus provide exactly the same conceptual analysis of our modal discourse as can the genuine realist, and arguably without the metaphysical price to pay.

A more serious difficulty for the modal fictionalist and the modal quasi-realist however is their inability to provide an adequate *semantics* for our modal discourse. That is to say, neither account can provide a fully extensional possible worlds semantics for our modal discourse whereby we can determine the validity of ordinary modal arguments within the resources of our standard, first-order methods. For the modal fictionalist and the modal quasi-realist – who both advocate a modal anti-realism grounded upon semantic considerations – the difficulty in satisfying this requirement lies in maintaining a semantics that is both fully extensional and appropriately normative, depending upon the type of semantic consideration in play; or in other words, by explicitly challenging some aspect of our familiar (realist) semantics, the modal fictionalist and the modal quasi-realist face a tension between getting the *content* of our modal statements right and getting the *inferential relationships* of our modal statements right. By contrast, committed modal agnosticism – since it specifically avoids making any semantic intervention – can meet both of these desiderata with the minimum of fuss.

For the modal fictionalist, who attempts to provide a re-interpretation of our modal discourse, it is the first of these considerations that is at issue – delivering a fully extensional semantics. The problem, as presented by Divers (1995), is as follows. According to the modal fictionalist, modal statements are to be analysed in terms of a fiction about possible worlds, rather than in terms of the possible worlds themselves. This is in fact a two-stage process: the modal statement is first rendered into a statement about possible worlds; then the possible worlds statement is understood as a mere fiction. An ordinary modal argument then, of the form

$$P_1, P_2, \ldots P_n \text{ therefore } C$$

is first translated into its possible worlds paraphrase:

$$P_1^*, P_2^*, \ldots P_n^* \text{ therefore } C^*$$

and then an 'according to the fiction' operator is prefixed to each premise and conclusion:

$$F(P_1^*), F(P_2^*), \ldots F(P_n^*) \text{ therefore } F(C^*)$$

The problem is simply that the fictionalist operator is a sentential operator that generates an intensional context (cf. Lewis, 1978): one cannot substitute co-extensional terms within its scope *salve veritate*. Take Divers' (1995: 85) example: the proposition 'there is a world with red dragons' is false according to the modal fictionalist, since according to him there are no possible worlds besides the actual world (at which there are no red dragons). Similarly, the proposition '$1 = 0$' is also false according to the modal fictionalist, although for somewhat more straightforward reasons. However, although '*according to the fiction*, $1 = 0$' remains false for the modal fictionalist on any sensible choice of modal fiction, the proposition '*according to the fiction*, there is a world with red dragons' is exactly the sort of claim that should come out as true. The fictionalist operator is thus an intentional operator: the truth-value of a proposition under the scope of the fictionalist operator is not truth-functional; prefixing the fictionalist operator does not have a determinate effect on the truth-value of that proposition. And as with other intensional operators – for example, believes that, desires that, doubts that, and so on – one is unable to apply a fully extensional logic to evaluate such notions.

The only recourse here would be to understand the fictionalist operator somewhat differently, such that it applies to an argument *as a whole*, rather than to each individual premise and conclusion. On this reading, the final formulation of our modal argument above would be of the form

$$F(P_1^*, P_2^*, \ldots P_n^* \text{ therefore } C^*)$$

This would certainly allow the application of a fully extensional first-order logic *within* the scope of the fictionalist operator, and would therefore appear to avoid the problem of intensionality. Unfortunately though, such a strategy seems illegitimate by modal fictionalist standards. According to the modal fictionalist, there are no possible worlds besides the actual world. It follows then that for the modal fictionalist, the argument within the fictionalist operator seriously misrepresents the modal argument to be evaluated (by reading the premises to be about possible worlds), and moreover, makes most of the premises of the modal argument simply false (again, by reading them to be about possible worlds). It is hard to see then how such an evaluation is of any use or interest to the modal fictionalist: by the time he comes to apply his fictionalist operator, the argument under consideration would already have been rejected as nonsense. As Divers (1995: 85–86) puts it, the modal fictionalist is thus caught in a dilemma between giving his modal arguments a credible semantic structure at the expense of extensionality, or capturing extensionality at the expense of interpreting our modal statements in a way that neither preserves their truth-value nor accurately represents their semantic structure: a dilemma between a non-extensional semantics or 'an extensional non-semantics'.[18]

The modal fictionalist attempted to rehabilitate our possible worlds discourse by construing it, not as making claims about other possible worlds, but rather as making claims about a *story* about other possible worlds. Such an approach promised to avoid the metaphysical excesses of modal realism by rejecting the existence of any non-actual entities, while still legitimising a suitable possible worlds semantics by taking our existential commitments to be true according to a certain fiction. Unfortunately, although the introduction of a fictionalist operator does generate a sufficient metaphysical gap between the modal fictionalist and an ontologically vast pluraverse, it also generates a sufficient semantic gap between the modal fictionalist and the possible worlds discourse he desires by rendering such discourse irreparably intensional. Moreover, since modal fictionalism also suffers from its inability to provide an adequate *actual* fiction in which to couch his account, the position has little to recommend it.

For the modal quasi-realist, who attempts to provide an anti-realist justification for our modal discourse rather than an anti-realist re-interpretation, it is the second of our earlier considerations that is of issue – delivering an appropriately normative semantics. The central issue here is the so-called Frege-Geach problem: although the quasi-realist may be able to give us a satisfactory account of how it is that we can come to assert simple modal statements like 'it is possible that p', our modal discourse consists of a variety of much more

complex, linguistic constructions, for example, embedded modal statements, complex molecular sentences containing modal assertions and
so forth. If quasi-realism is to successfully explicate our modal discourse,
it must also account for these more sophisticated locutions: '*I wish it were
the case that* it is possible that p', '*if* it is possible that p, *then Bas will be
happy*' and so on. Yet as many have pointed out (see, e.g., Geach, 1960;
1965), when modal statements appear in these complex constructions,
they appear to express *propositions* rather than the attitudes with which
the quasi-realist identifies them: when I *wish that* it is possible that p,
I do not appear to be expressing anything about my conceptual limitations; and when asserting that Bas will be happy *if* it is possible that p,
I leave the issue of whether or not p is possible entirely undecided, and
again no attitude is expressed.

This is of course a venerable objection to quasi-realism, and one to
which much ink has been spilt in defence, and so only a brief summary
is presented here. Blackburn's (1988) general strategy for accommodating these apparently propositional contexts involves the notion of a
conditional commitment, best illustrated with respect to disjunctive modal
statements. When I assert 'either p is possible or q is possible', I do not
appear to assert any particular attitude towards either disjunct; however,
what I do assert is a commitment to a certain pattern of reasoning – to be
able to make something of ways of thought involving p if I cannot make
anything of ways of thought involving q, and *vice versa*. And as Rosen
(1998: 389–393) notes, this idea of a commitment to various patterns of
attitudes provides an effective general strategy for any arbitrary complex
molecular sentence. Firstly, one gives a quasi-realist treatment of the
basic atomic sentences making up the molecular sentence – 'it is possible
that p' – in the manner sketched above. Secondly, one gives an account
of negating a basic modal sentence, being *unable* to make anything of
ways of thought involving p. Finally, since the standard set of propositional connectives {¬, &, ∨} is functionally complete, we can rewrite
any arbitrary complex molecular sentence φ in its disjunctive normal
form. Each disjunct will then consist of a set of basic statements or their
negation, and the attitude expressed by φ will be a commitment to take
up the attitude expressed by any one set of basic statements should one
come to reject the attitudes expressed by the other, exclusive, disjuncts.
To return to an earlier example then, the apparently propositional context expressed by '*if* it is possible that p, *then Bas will be happy*' can be
rewritten as '*either* (it is possible that p and Bas is happy) *or* (it is not
the case that it is possible that p and Bas is happy) *or* (it is not the case
that it is possible that p and Bas is unhappy)'. Consequently, this can be

accounted for quasi-realistically as the commitment to *either* (being able to make something of ways of thought involving p, and Bas is happy) *or* (being unable to make something of ways of thought involving p, and Bas is happy) *or* (being unable to make something of ways of thought involving p, and Bas is unhappy).

The residual difficulty with this approach however concerns the notion of logical consistency. Take a simple *modus ponens* with modal statements as premises:

1. If it is possible that p, then q;
2. It is possible that p;
3. Therefore q.

The argument is clearly valid; and anyone who endorses (1) and (2) but did not endorse (3) would be rightly held to have violated some logical norm. But according to the quasi-realist, (1) simply asserts a commitment to the truth of q if I can make something of ways of thought involving p. It follows then, as Wright (1988: 33) notes, that if the quasi-realist endorses (1) and (2) without endorsing (3), he violates his commitment and is thereby *pragmatically* inconsistent (and depending upon the context of these utterances, perhaps *morally* inconsistent insofar as his actions may be likened to breaking a promise); but there seems to be no sense in which the quasi-realist can capture the logical, cognitive inconsistency that we feel is lurking here.

The problem then is this. In order for the modal quasi-realist to provide an adequate semantics for our modal discourse, he must rehabilitate both the *extensional* and the *normative* structure of a standard, Kripke-style possible worlds framework. We have supposed for the sake of argument that the quasi-realist can provide the required extensionality insofar as he can successfully imitate the propositional form of a possible worlds discourse (justified as a series of pragmatically motivated refinements upon an underlying practice of expressing our own cognitive limitations). But in granting the success of this programme, it seems that the quasi-realist cannot capture the inferential structure that holds between the statements of our modal discourse, since the story he tells fails to accommodate the *logical* inconsistency of asserting a contradiction.

This leaves the quasi-realist with two options. The first is to bite the bullet, and simply to maintain that our standards of logical normativity are ultimately reducible to our standards of moral normativity. The thought would be that a fully consistent quasi-realist would also offer a fully projectivist treatment of our logical discourse in much the same way as he would of our modal discourse, that is, explain the 'hardness of the logical must' in terms of our various (non-logical) commitments. But even leaving aside any issues concerning the plausibility of such an account of our logical notions, the problem with such a response is that it just makes modal quasi-realism too expensive. In order to provide an anti-realist account of our possible worlds discourse that can both give a satisfactory conceptual analysis of our modal discourse and capture the structure of a fully extensional semantics for that discourse, the modal quasi-realist must first offer a substantial – and in many ways quite revisionary – account of the nature of our logic. Given then that my concerns in this section are simply to compare the relative pros and cons of alternative anti-realist accounts of possible worlds, such a response is not an attractive prospect for the quasi-realist.

The other option, and the one that seems to be endorsed by Blackburn, is to attempt to capture a notion of logical consistency in the above inference by appeal to some form of deontic logic, along the lines of Hintikka (1969). The idea is that a set of commitments can be deemed logically inconsistent insofar as there is no possible world where they can all be jointly satisfied. Again I wish to leave to one side any issues concerning the plausibility of such a response.[19] For just as with the first response, there is a serious difficulty with the modal quasi-realist offering such a reply regardless of its plausibility. The problem here is how exactly one is to understand the notion of a possible world introduced by our system of deontic inferences; and here it looks as if the modal quasi-realist has simply arrived back at where he started. If one goes quasi-realist about *these* possible worlds too, then Wright's objection will simply repeat at the meta-level; and if one proposes a non-quasi-realist account of these worlds, then one loses any benefits that one might have acquired in offering the quasi-realist account in the first place.

It would appear then that neither modal fictionalism nor modal quasi-realism can provide the anti-realist with an adequate philosophy of modality. Modal fictionalism cannot provide a conceptual analysis of our modal discourse, since it must presuppose such notions in order to ground everything we might want to say; modal quasi-realism arguably faces the same problem, although its prospects do look more promising, More seriously though, neither modal fictionalism nor modal quasi-realism can rehabilitate the extensional, model-theoretic logic for our modal discourse that I take to be the major desiderata of any adequate account: either because they fail to capture extensionality (fictionalism) or because they fail to capture the normativity integral to such a logic (quasi-realism). By contrast, the position that I have called committed modal agnosticism can provide both. According to the committed modal agnostic, our modal discourse is to be understood as being about, and being made true by, other possible worlds. Unlike the modal fictionalist and the quasi-realist then, the committed modal agnostic does not need to introduce additional semantic complication into his analysis: neither intensional operators nor a logic of attitudes. The committed modal agnostic can thus provide exactly the same *semantic* analysis of our modal discourse as can the genuine realist; and again, all this (arguably) without the metaphysical price to pay.

Committed modal agnosticism does however face its own set of difficulties; and given that the position is an anti-realist approach to possible worlds grounded upon considerations of epistemology, it is perhaps unsurprising to note that the main difficulties besetting the position are epistemological. Specifically, the main problem facing the committed modal agnostic is the serious degree of *modal scepticism* that his position seems to entail. For the committed modal agnostic, our modal discourse is to be understood as making claims about what is and is not the case at other possible worlds, and that such claims are determinately true or false depending upon the nature of these other worlds. However, the committed modal agnostic also refuses to assert whether or not these other worlds really do exist; consequently, he is unable to assert the truth-value of much of our modal discourse.

Of course, there will be *some* modal statements which the committed modal agnostic can assert the truth-value of. As Divers (2004; and see above) has noted, the modal agnostic believes in the existence of one possible world – the actual world – and insofar as the truth-value of a modal statement can be known solely on the basis of this fact, some modal statements (primarily negative existential statements) will be unproblematic. Similarly, although somewhat more restrictively, I have argued that the *constructive empiricist* committed modal agnostic can know the truth-value of any modal statement that can be known solely on the basis of *observable* facts about the actual world. Nevertheless, for the vast majority of modal statements that he might make, the committed modal agnostic can only accept the truth-value of that statement. However, it remains to be seen to what extent the modal epistemology forced upon the committed modal agnostic is in fact a vice rather than a virtue.

4.3.4   A short note on mathematical nominalism

We have seen then how the acceptance solution provides a straight-forward response to the objections raised by Musgrave and Ladyman; we have also seen how the philosophical account of modality resulting from the acceptance solution is a workable anti-realist proposal, which may even enjoy some significant overall advantages to its rivals. Our final task then is to investigate the plausibility of offering the same style of response to Rosen's concerns over the existence of abstract mathematical objects. The problem of course is structurally identical to the issues surrounding the unobservability of unobservable phenomena, and of the truth-conditions of the constructive empiricist's counterfactual conditionals: Rosen argues that in order for the constructive empiricist to even state what his view regarding the aim of science amounts to, he must believe in the existence of abstract mathematical objects (since his view explicitly attributes a certain property to the models of our scientific theories, and models are abstract mathematical objects); yet the constructive empiricist's view is *precisely* the claim that we need not believe in the existence of such entities in order to make sense of scientific practice. And just as with the issues of unobservability and modality, we have seen the structurally identical response offered on the constructive empiricist's behalf: that the problematic beliefs in question are in fact reducible to a weaker, but internally consistent, set of beliefs (in this case, the belief that a particular story that we can tell about the existence of abstract mathematical objects is a conservative extension of our non-mathematical discourse); and that whatever deficit is left between what the constructive empiricist can believe and what he needs to believe is to be patched up within the minimal context of van Fraassen's permissive conception of rationality. Yet the problem of abstract mathematical objects was shown to be both the climax and the denouement of this project – for since the deficit left by the mathematical fictionalist strategy was precisely those beliefs concerning logical consistency upon which the epistemic voluntarist must rely, no such option is available.

The acceptance solution to the problem of abstract mathematical existence must by now be fairly clear: the constructive empiricist should concede to Rosen that ne needs to *use* various claims about the existence of abstract mathematical objects, but that in the context of his principled distinction between acceptance and belief, it does not follow that he thereby needs to *believe* them – one then proceeds to elaborate upon the substantial epistemological commitment constitutive of the attitude of acceptance; the analogy between accepting straightforward claims about concrete unobservable phenomena for the purposes of subsequently using such claims in one's predictions and explanations and accepting claims about abstract mathematical objects and subsequently using them in the articulation of one's view of science; and the argument that, since these claims regarding the existence of abstract mathematical objects function somewhat like framework constitutive principles, they may not even be legitimate candidates for belief anyway.

All of this should be perfectly predictable, and I shall offer no further arguments for the acceptance solution in general in the context of the philosophy of mathematics. It is also beyond the scope of this book to discuss the prospects for, and the consequences of, an explicitly constructive empiricist account of abstract mathematical objects. This is because the crucial issue concerning the acceptance solution for the problem of mathematics is not so much with the overall philosophy of mathematics to which the constructive empiricist is thereby committed, but with some of the *modal* consequences of this strategy. In his original paper, Rosen (1994: 167–168) identifies something like the solution presented here: that the constructive empiricist should simply remain agnostic about the existence of abstract mathematical objects – although it should be noted that Rosen does not make the connection between such a response and the constructive empiricist's distinction between acceptance and belief. His conclusion however is that this sort of strategy cannot work. He argues that if the constructive empiricist is going to be agnostic about abstract mathematical objects, then 'to accept [a theory] is to believe that the world is such that *if there were such a thing as* [the theory], *it would be empirically adequate*' (1994: 167 [original emphasis]). It follows then that if the constructive empiricist is going to adopt a mathematically agnostic strategy, then in order to state what constructive empiricism amounts to, he must assert a counterfactual – which is just to say that the immediate consequence of mathematical agnosticism is that the constructive empiricist's statement of his position must be essentially modal in nature.

But this, according to Rosen, raises a serious problem. He argues that

> [t]he trouble with all of this from van Fraassen's perspective is modality itself... there is reason to think that the constructive empiricist is committed to agnosticism about a range of modal facts. The counterfactuals that constitute the entities of belief on the present proposal arguably fall into that class. And if that is right then whatever its merits, the proposal is not available to van Fraassen.
>
> (1994: 168)

So, according to Rosen, regardless of the other issues surrounding mathematical agnosticism, the most important consequence of such a view is that it commits the constructive empiricist to a modal statement of his position; but since the constructive empiricist must also be agnostic about this modal statement, it is a statement he cannot believe. Mathematical agnosticism therefore entails the unfortunate consequence that the constructive empiricist can no longer believe what his position is supposed to be.

Fortunately, however, in the light of the preceding discussion regarding modal agnosticism, this is a conclusion that we can happily resist. For not only can the constructive empiricist accept such putatively problematic modal statements, our lengthy examination of modal anti-realism in general has shown how such a strategy may also be independently attractive. The real consequences of the constructive empiricist's mathematical agnosticism are therefore just as much modal as they are mathematical, and in that respect have already been resolved.

4.4 Summary

Constructive empiricism consists of two fundamental distinctions: the distinction between observable and unobservable phenomena that sets the limits of what one needs to believe in order to make sense of scientific practice; and the distinction between acceptance and belief that underscores the constructive empiricist's attitude towards those parts of his scientific theories that he does not need to believe but may still need to use. In this chapter, I have argued that it is the latter of these two distinctions that is in fact the more important – for in the light of the problem of internal coherence, and the complete inability of van Fraassen's epistemic voluntarism in resolving that problem, it is the distinction between acceptance and belief that ultimately grounds the constructive empiricist's allegedly more fundamental distinction between observable and unobservable phenomena.

Fortunately, we have also seen in this chapter that the distinction between acceptance and belief is in fact the most philosophically robust component of the constructive empiricist's position: although widely derided, there have been few cogent objections actually raised against this distinction; and of the few that have been raised, we have seen them all to rest upon both a highly specific philosophy of mind and an unrealistically simplistic conception of belief. Moreover, I have argued that a proper appreciation of the individuation of intentional states – and in particular, the issue as to whether or not a particular intentional state can be entered into voluntarily – provides ample resources for contrasting these different attitudes. Perhaps most importantly, I have also argued that something like the constructive empiricist's distinction between acceptance and belief is a ubiquitous element of the philosophy of science – both in its importance to a spectrum of positions within the scientific realism debate and in its importance in making sense of a broad range of features of our contemporary scientific practice. To rest his position upon the distinction between the different attitudes that we can hold towards a proposition or theory, rather than upon the distinction between the different standards of rationality that we can appeal to in evaluating an epistemological policy, is very much for the constructive empiricist to play to his strengths.

But the appeal to the distinction between acceptance and belief not only resolves the constructive empiricist's problem of internal coherence (which, in the light of the rather inconclusive criticisms raised concerning the justification of the constructive empiricist's position, was by far the most substantial difficulty faced), it also opens up some promising new terrain. In responding to Ladyman's objection, we have explored a distinctively constructive empiricist option in the philosophy of modality – a form of modal agnosticism backed up with a substantive distinction between what one finds credible and what one finds pragmatically expedient. And in the philosophy of science, we have now tied the fate of empiricism to the wealth of different attitudes that had so far remained largely implicit in our understanding of scientific practice. Constructive empiricism may ultimately fail to attract any further adherents; but in giving it what I believe is the best reformulation and defence that I can, I hope at least that the next salvo of criticisms will begin to take seriously some of the more interesting facets and dimensions of empiricism that have been exploited in the present discussion.

Conclusion: What Is This Thing Called 'Constructive Empiricism'?

Since its inception, constructive empiricism has sought to challenge the received wisdom that is scientific realism – the view that in order to make best sense of contemporary scientific practice, one must assume that our scientific theories offer literally true depictions of the unobservable reaches of reality. Yet the precise nature of this opposition has been frequently misunderstood. For while it is generally recognised that constructive empiricism is in essence a philosophical view regarding the epistemology of science – as opposed to, say, a semantic view regarding the language of science – it is usually supposed that this must mean that constructive empiricism is in essence a *sceptical* view regarding the epistemology of science. The challenge posed by the constructive empiricist is however far more radical, and far more interesting, than such a caricature allows. For the constructive empiricist doesn't simply challenge the conclusions drawn by the scientific realist, he challenges the epistemological framework in which such arguments proceed: not with whether or not the truth of our scientific theories provides the best explanation for their success, but with whether or not our best explanations have any epistemological relevance; not with whether or not our methods of instrumental detection are just as reliable as our unaided observations, but with whether or not questions of reliability are the most important factor in belief revision; not with whether or not we can make extrapolative inferences about the future, but with whether or not we have to.

Constructive empiricism then is in essence a meta-philosophical view about how one is to proceed in the philosophy of science. It is this that I take to be the most important and valuable feature of the position, and it is this that I have attempted to explore and elaborate upon in this book. For while I whole-heartedly agree with van Fraassen that one cannot attend to the scientific realism debate without thereby attending to one's epistemological presuppositions, and moreover that to endorse an alternative to scientific realism regarding the *aim* of science is thereby to endorse an alternative to scientific realism regarding the *rationality* of science, I do not believe that his own epistemic voluntarism offers the most satisfactory approach to this issue.

This is not because I reject epistemic voluntarism *per se* (although I have indeed voiced various misgivings about such a minimalist conception of rationality throughout this book). Rather, it strikes me that the issue of epistemic voluntarism – more specifically, the issue of whether or not scientific rationality should be understood as a matter of obligation or of permission – is somewhat orthogonal to our primary concern regarding the methodology of the philosophy of science. There are two main reasons for my conclusion here. The first is that epistemic voluntarism is *unnecessary* for the articulation and defence of constructive empiricism. The primary motivation for the constructive empiricist to endorse van Fraassen's minimalist conception of rationality is as a response to those objections to the distinction between observable and unobservable phenomena that seek to show some justificatory equivalence between what our scientific theories say regarding what we can see and what we can detect – for example, that our unobservable ontologies are just as immediately given, or just as vulnerable to revision, as our observable ontologies. Such objections clearly presuppose a traditional epistemological framework in which questions of evidence outweigh any other considerations for belief revision, and in which one is always obliged to make the same sorts of inference when faced with the same sorts of data; consequently, adopting an epistemological framework in which one allows one's values to compete with one's evidence, and where inference is always optional, easily disarms such objections. Yet to adopt epistemic voluntarism as a response to the scientific realist's ringing endorsement of the electron microscope is to take the proverbial sledgehammer to crack the proverbial nut: not only do such objections fail to draw the necessarily quantitative conclusions from their purely qualitative premises, many of them simply beg the question even taken on their own terms. Quite simply then, one does not *need* to be an epistemic voluntarist in order to be a constructive empiricist.

Secondly, and more importantly, epistemic voluntarism is also *insufficient* for the articulation and defence of constructive empiricism. This is the conclusion to be drawn from the various problems of internal coherence advanced by Musgrave, Ladyman and Rosen, respectively. In order to draw his basic distinctions between what he needs to believe in order to make sense of scientific practice and that which he does not need to believe, the constructive empiricist finds that he must in fact believe some of those claims that he wishes to classify as unnecessary. The voluntarist response to this problem has been to endorse a restricted (and thereby internally coherent) class of beliefs, and to attempt to make up any doxastic deficit with an amended epistemic policy that depends for its legitimacy upon endorsing an epistemic framework that demands very little in the way of philosophical justification. Yet even van Fraassen's minimalist conception of rationality depends upon a basic account of logical consistency and probabilistic coherence – and when these themselves are called into question, the voluntarist is rendered helpless.

None of this of course is to say that the constructive empiricist should *not* be an epistemic voluntarist – this will be largely determined by how one is taken by van Fraassen's additional arguments regarding the positive constraint of diachronic probabilistic coherence, the negative assessment of ampliative inference and, perhaps most importantly, his re-conceptualisation of empiricism in terms of an epistemic stance (although I have also expressed some misgivings about these too). The point is rather that while the constructive empiricist may or may not wish to be an epistemic voluntarist, he must also give some account of how he is to render his position internally coherent – either in *addition* to epistemic voluntarism (which we have seen to be unable to resolve this difficulty), or more simply as an *alternative* to epistemic voluntarism (which we have seen to be unnecessary for constructive empiricism).

The appropriate epistemological framework for the articulation and defence of constructive empiricism is however already present in van Fraassen's original formulation. For in addition to the much maligned distinction between observable and unobservable phenomena, constructive empiricism also depends upon a distinction between what one *believes* and what one *accepts* – a distinction between the different attitudes that one can hold towards a proposition or theory, for broadly evidential and pragmatic reasons, respectively. Suitably elaborated, this distinction provides a highly satisfactory solution to the problem of internal coherence: for if the constructive empiricist can be sufficiently committed to those putatively problematic claims that underlie his other fundamental distinctions by merely accepting them, then it is of little consequence if he cannot also believe them. Moreover, a principled distinction between those elements of a scientific theory that one believes and those elements that one endorses for their theoretical utility provides a healthy framework for understanding those aspects of scientific practice that would otherwise seem anomalous or even contra-dictory: the simultaneous endorsement of mutually inconsistent models or theories; the continuing endorsement of a scientific theory in the face of falsifying data; and the use of incompatible assumptions in the derivation of a new result.

Most importantly however – at least for our present purposes – this reorientation towards the different *attitudes* involved in our understand-ing of scientific practice, as opposed to the different *standards of rational-ity*, remains firmly in the spirit of what I take constructive empiricism to be. To defend an empiricist view of the aim of science is to defend an empiricist view of the appropriate epistemological framework in which to pursue the philosophy of science. It is a framework that focuses not upon whether or not we have greater justification in believing what our scientific theories have to say about observable phenomena than they do about unobservable phenomena (although it may also include this), nor upon whether or not we are compelled to believe that our best explanations are likely to be true (although it may also have something to say about this too), but ultimately upon whether or not the question of what we believe and disbelieve provides a sufficiently fine-grained understanding of contemporary scientific practice.

Notes

1   Arguments Concerning Constructive Empiricism

1. I owe this tripartite distinction to Psillos (1999: xix).
2. This would appear to be the view of at least some of those involved with the so-called sociology of scientific knowledge; see, for example, Bloor (1976).
3. By far the best discussion of the motivations, weaknesses and possible refinements of this view is still Carnap (1936, 1937). See also Psillos (1999: Ch. 1).
4. This view is mainly associated with Ernst Mach (see, e.g., his 1893; 1910). For critical discussion, see Nagel (1950), Hempel (1958) and Psillos (1999: Ch. 2).
5. My favourite is the indispensable Psillos (1999: Ch. 9).
6. A similarly rhetorical argument can be found in Chihara and Chihara (1993) – here the authors consider the example of a tiny mite, which with the aid of a microscope we can determine to be carried around on eight even tinier legs. Since we cannot observe these legs directly, the constructive empiricist would presumably withhold belief in their existence; yet when these legs are removed (at least, according to our microscopic observations), the mite is directly observed to become stationary; and when these legs are removed from only one side of its body (again, according to our microscopic observations), the mite is directly observed to wander about in a circle (ibid.: 654–655). As with Hacking's example, we seem forced to conclude that *contra* the constructive empiricist, those detections we make through our scientific instruments are just as well supported as those observations we make with the naked eye.
7. There is another difficulty pertaining to the constructive empiricist's attitude towards modal matters, closely related to the problem raised by Musgrave above, that is, with whether or not the constructive empiricist's views regarding the aim of science are in fact compatible with the attitudes he must hold towards the modal consequences of our scientific theories in order to maintain such a view. This is discussed in detail in Chapter 3.
8. In addition to this somewhat negative characterisation, van Fraassen's epistemic voluntarism also contains a positive thesis regarding the nature of epistemic judgements – roughly speaking, that to assert one's beliefs (or one's subjective probability in a particular proposition, etc.) is not so much to express an autobiographical fact about one's own mental state as it is to undertake a commitment towards a particular course of action. This feature is discussed in detail in Chapter 2.

2   Epistemic Voluntarism: Rationality, Inference and Empiricism

1. As van Fraassen notes, the Reflection Principle was independently proposed by Goldstein (1983); however, as my concern in this chapter is with van

Fraassen's epistemic voluntarism in general – and the role the Reflection Principle plays in motivating this framework – the discussion here will be limited to van Fraassen's articulation and defence of the principle.

2. My expectation for the first wager $= [(0.2 \times 1) + (0.8 \times 0)] - 0.2 = 0$; my expectation for the second wager $= [(0.6 \times 0.5) + (0.4 \times 0)] - 0.3 = 0$; my expectation for the third wager $= [(0.4 \times 0.5) + (0.6 \times 0)] - 0.2 = 0$.

3. Maher (1992: 124–126) offers a similar argument against using a diachronic Dutch-book strategy in order to motivate the Reflection Principle. He argues that there is no reason to suppose that an epistemic agent who violates the Reflection Principle will be vulnerable to a diachronic Dutch-book (and hence, there is no reason to suppose that an epistemic agent who violates the Reflection Principle is thereby irrational) since there's no reason to suppose that such an agent would ever *accept* the betting strategy offered to him in the first place: essentially, since anyone offered such a series of wagers would be just as able to calculate their overall (negative) payoff as the bookie offering them, any such agent would realise that they would maximise their utility by simply declining to bet (even though each wager considered individually would seem fair). The analogy Maher draws here is with whether or not a rational agent would take a fair bet on the likelihood of rain, in the full knowledge that any winnings would be subject to a hefty gambling tax; he notes that 'Bayesians, when careful, have always said that in making decisions, one needs to look ahead and take into account what may happen in the future' (1992: 125; cf. Levi, 1987: 204–205). The point then is that unless one *presupposes* the sort of diachronic constraint imposed by the Reflection Principle, there is no reason to expect an epistemic agent to consider his potential vulnerability to diachronic incoherence as a matter of irrationality.

4. For a more concrete example, and as our earlier discussion of the Reflection Principle showed, the epistemic voluntarist is at least committed to an objective account of the calibration of an agent's beliefs (e.g., in terms of announced probabilities of rain, and the actual frequency of rain), which would be utterly undermined by the familiar radically sceptical hypotheses. For more on the distinction between epistemic voluntarism and radical scepticism, see van Fraassen (1989: 176–182; 2000: 277–279).

5. For a stimulating discussion of this interpretation of Peirce, see Misak (2004).

6. The following argument does not depend on any particular choice of unit – indeed, for any putatively perfectly calibrated method of inference, there will be an infinite number of ways in which it fails to accommodate every computable subsequence as there will be an infinite number of different units (years, months, weeks, milliseconds, . . .) in which to construct the argument.

7. Good discussions of this approach to the justification of induction can be found in van Cleve (1984), Mellor (1988) and Lipton (2000).

8. An explanation is more likely than another if it has a greater chance of being true, and an explanation is more lovely than another if it provides a better explanation. For the definitive discussion of these competing virtues, and on how to characterise an inference to the best explanation in general, see Lipton (2004).

9. We may, for example, argue that evolutionary pressures will have selected for creatures that perform largely truth-tropic comparisons of hypotheses; or if one prefers, that benevolent design will have guaranteed it. But the

divine plan is beyond our ken; and reproductive success has an equally unfathomable correlation with representational accuracy.

10. For example, if our accepted background theories explicitly reject the possibility of a certain type of entity or process, any potential explanation that posits such things will thereby be rendered very unlikely, regardless of its other theoretical virtues.

11. This of course is related to the Frame Problem in the philosophy of artificial intelligence; see Dennett (1978) and Fodor (1983).

12. Similar presentations of this argument can also be found in van Fraassen (1994; 1995b). The latter paper explores the options for an empirical (i.e., naturalised) statement of the empiricist's core position in greater detail.

13. Suppose that the metaphysically loaded alternative to the empiricist's central dictum *was* open to empirical investigation, that we could somehow test the reliability of *a priori* reasoning. In this case, Jauernig's amended policy would not render such claims inadmissible. But if it really were the case that the metaphysician was suggesting an alternative that admitted of empirical investigation, it is hard to see how this would be an *alternative* to empiricism in any philosophically substantial sense (Jauernig, 2007: 280–281).

14. The structure of this objection is of course closely related to the argument we encountered in the discussion of the Reflection Principle: that just as understanding our epistemic judgements as expressing a commitment to a certain epistemic policy does not rule out the possibility of also coming to hold various autobiographical beliefs *about* these commitments, neither does understanding empiricism as a certain epistemic policy rule out the possibility of also coming to hold various autobiographical beliefs *about* this policy. See Section 2.2.4.

15. Similar worries are found in Jauernig (2007) and Ho (2007).

16. In particular, Nagel (2000) has argued that while the empiricist may well be able to afford to experience a privileged position within his epistemology, he must still make an ineliminable appeal to metaphysical speculation in order to understand what 'experience' is; this issue is further discussion in Ladyman (2007) and Chakravartty (2007).

17. Recall the discussion in Section 2.3.4 regarding the weaknesses of a purely probabilistic approach to epistemology, and how various non-structural (in this case, explanatory) considerations are required in order to help us determine our prior probabilities, our expectations of a hypothesis conditional on a piece of evidence, or even when something counts as a piece of evidence in the first place.

3   Against Epistemic Voluntarism: Musgrave, Modality and Mathematics

1. See, for example, Kukla (1998: 138–139) and Muller (2004); Dicken and Lipton (2006) discuss several possible interpretations.

2. Another prominent proposal is that such knowledge is based simply upon considerations of some kind of conceivability (see, e.g., Yablo, 1993, or more recently Chalmers, 2002; for a survey of various other strategies with respect to modal epistemology, see Hawthorne, 1996). The point of course is that

the various mechanisms proposed for securing our modal epistemology are all of a very different kind from the predominantly causal and/or abductive reasoning that seems to characterise most of our knowledge of the actual world.

3. Since my aim in this section is merely to contrast our epistemological access to actual unobservable phenomena with our epistemological access to merely possible phenomena, I shall for the purposes of clarity restrict my discussion to a Lewisian modal realist account. My argument however applies *mutatis mutandis* to other (ersatz) modal realist accounts of non-actual phenomena.

4. Ladyman dubs the combination of constructive empiricism and modal realism 'modal empiricism', after the position sketched by Giere (1985: 83).

5. Again, recall the discussion of Churchland's objections in Chapter 1.

6. There is a serious and rather fundamental worry with all of this, since as my exposition of van Fraassen's position makes clear, one must make frequent reference to the *consequences* of the constructive empiricist's accepted scientific theories. This then raises the worry as to whether or not van Fraassen's deflationary account of modality is simply circular, since the notion of consequence is of course a modal notion. However, I shall not discuss this issue here, since (a) it is not clear to me what van Fraassen's views regarding this matter are, (b) I shall eventually reject van Fraassen's approach on completely independent grounds and (c) problems concerning the constructive empiricist's general account of meta-logical notions – such as 'consequence' and 'consistency' – will be discussed in detail in the next section.

7. Actually, it is important to note that some of van Fraassen's pronouncements on the matter of modality are somewhat less than clear, and in fact rather difficult to reconcile with one another; although I believe the interpretation presented here to be correct, see Ladyman (2000) for a fuller discussion of the various possible interpretations of van Fraassen's position.

8. That our modal discourse is to be understood within a two-dimensional framework such as this is to allow a sense in which, say, our claims of nomological necessity can be understood as only *contingently* necessary; or in other words, that our 'laws' could have been different. Without this extra dimension of evaluation, all speakers at all possible worlds will pick out the same regularities when they speak of 'the laws of nature', and claims of nomological necessity will therefore become simply tautologous. For a fuller discussion of his semantic analysis of modality, see van Fraassen (1977b, 1978, 1981). For more on the two-dimensional semantic framework to which van Fraassen originally appeals, see Stalnaker (1972; 1974). For a comprehensive recent survey of two-dimensional semantics, see Garcia-Carpintero and Macia (2004).

9. We can of course make *some* progress on this issue by noting that the set of models in which the constructive empiricist will be interested with respect to those counterfactual conditionals relevant to drawing his distinction between observable and unobservable phenomena will be the set of models of his accepted theory of observability – presumably including some suitable sub-theory of the Faraday-Maxwell theory of electrodynamics, for example. For more details on this, see Muller (2005: 71–82). It should be noted however that although Muller develops this aspect of Monton and

van Fraassen's proposal at length, his formulation has absolutely nothing to say on the deeper – and far more important – problem concerning the similarity ordering of these models, which of course constitutes the main focus of this section.

10. It is important to note of course that the sort of laws that the constructive empiricist needs to appeal to in order to privilege one model of his accepted scientific theory over another will be the laws of his *other* accepted scientific theories (since, clearly, all of the models of a single theory will hold all of that theory's laws in common). So for example, when selecting the appropriate model of his theory of observability for evaluating counterfactual conditionals about what he would or would not be able to see, the sort of facts that the constructive empiricist needs to keep constant will be (among others) those that appear as laws in his theories regarding the constitution of moons, the physiology of dinosaurs and so forth.

11. Moreover, there may be something of an *ad hominem* lurking here, since Ladyman (Ladyman and Ross, 2007) pursues his philosophy of science against the backdrop of a highly naturalised metaphysics. Arguably then, there is something of a methodological tension to be found between his view that we should essentially allow contemporary physics to settle our metaphysical disputes and the metaphysically loaded criteria he brings to bear in his criticism of Monton and van Fraassen's deflationary account of counterfactual conditionals. The extent to which one finds the meta-linguistic approach unsatisfactory will be determined by the sort of pre-scientific cost-benefit analysis of ontological commitment that one brings to the debate, the sort of thing championed by David Lewis and supposedly renounced by the naturalistic metaphysician. Indeed, a purely philosophical debate over the attractive features of a theory of modality is something that floats quite freely of any naturalistic constraint; and no debate over the appropriate semantics for counterfactual conditionals is going to make any headway in a unified, naturalistic metaphysics. For more, see Dicken (2008).

12. More specifically, van Fraassen takes a scientific theory to consist of a set of Suppesian models, which are themselves some kind of abstract, set-theoretical object; for more, see van Fraassen (1980: 64–69; 1989: 217–232).

13. It should of course be noted that since its initial exposition, Field's programme has been subject to an absolute barrage of criticisms, all of which should give the constructive empiricist pause in his endorsement of such a strategy. It has been objected, for example, that our mathematical theories cannot be shown to be conservative *in principle*, because of the incompleteness of arithmetic (Shapiro, 1983); and that since Field's nominalisation strategy must presuppose an infinite number of space-time points (or other nominalistic surrogates), it is in fact just as ontologically profligate as the platonistic formulations it seeks to replace (Melia, 1998). It has also been objected that Field's general nominalisation strategy for the natural sciences must fail since quantum mechanics is irreducibly statistical and that one cannot find adequate nominalist surrogates for the probabilities involved (Malament, 1982). The most plausible response to this problem is due to Balaguer (1998), who has suggested that a nominalisation strategy could be carried out in terms of physically real propensities. Amusingly

enough, this is a response that Bueno himself has explicitly rejected – he argues (2003) that propensities are not nominalistically acceptable; that Balaguer's strategy constitutes an *interpretation*, rather than a nominalisation, of quantum mechanics; and perhaps most importantly for our purposes, that Balaguer's account is fundamentally inconsistent with van Fraassen's (1991) modal frequency interpretation of quantum mechanics. In fact, Bueno goes so far as to conclude 'that the nominalisation of [quantum mechanics] remains a major problem for the nominalist' (2003: 1435). Regardless then of what one makes of the other objections raised against Field, it seems that Bueno himself believes that mathematical fictionalism still requires a lot of substantial improvement. Consequently, Bueno's motivation for his proposed combination of constructive empiricism and mathematical fictionalism seems somewhat at odds with own assessment of its prospects.

14. Actually, (1) is due to Melia (1992: 38). Although Hazen presents several different examples of intuitively true modal claims that cannot be expressed with a primitive modal language, I shall, for the sake of clarity, restrict my focus to Melia's example. This, I hope, shall help to make explicit exactly what difficulties face a primitive modalist like Field – and by extension, any constructive empiricist who attempts to replicate his strategy. What follows however will of course apply *mutatis mutandis* for all of the other so-called Hazen sentences.

15. A discussion of various attempts to remedy the expressive incompleteness of the mathematical fictionalist's primitively modal account of consistency – all of them unsuccessful – is included as an appendix to this chapter.

16. A more recent – although even less compelling – attempt to provide deflationary translations of sentences such as (1) and (2) is due to Nolan (2002). His proposal is to introduce an ontologically neutral quantifier, along the lines of Routley (1980), and hence to quantify over objects (possible worlds, models, etc.) that are explicitly held not to exist. However, as was the case with Forbes' proposal, it is far from clear just how intelligible such a strategy is. Even leaving aside the radical revision such a proposal requires for our understanding of the quantifiers, and the nominalistically dubious appeal to second-order quantification that Nolan is forced to make, it must still be asked how these ontologically neutral quantifiers are to be understood. As with Forbes, it seems that the only way we are to understand the proposed paraphrase is in terms of a traditional possible worlds semantics, a comparison that Nolan himself invites (ibid.: 55). Consequently, it appears that Nolan's strategy is also parasitic upon the notion of quantifying over model-theoretic objects, and is thus similarly unacceptable to the mathematical fictionalist.

17. My eventual solution for the constructive empiricist – taken up in detail in Chapter 4 – can also be glossed in terms of finding some way to use model-theoretic reasoning without endorsing its ontological commitments. However, whereas my proposal is in essence an *epistemological* strategy that develops the constructive empiricist's distinction between acceptance and belief, it should be clear that the proposal under discussion here is still a *deflationary* strategy in that it attempts to reduce the belief in the existence of abstract mathematical objects to the belief about the logical consequences

of various aspects of our mathematical discourse. Consequently, such an approach remains firmly within the general methodology that is under critique in this chapter.

4 On the Nature and Norms of Acceptance and Belief

1. For example, in his original paper, Melchert diagnoses the constructive empiricist's mistake as attempting to occupy an untenable middle ground between full-blown scientific realism and traditional (syntactic) instrumentalism through endorsing an epistemological distinction between acceptance and belief rather than the familiar strategy of semantic reinterpretation; and it is because no such middle ground can be consistently maintained that this epistemological distinction appears bogus. Thus while Melchert may well have perfectly good reasons for rejecting constructive empiricism in general, it is clear that it is for *these* reasons that he rejects the distinction between acceptance and belief, rather than the other way around. Insofar then as we are looking for an actual *argument* against the distinction between acceptance and belief, we appear then to be left with the aforementioned rhetorical flourishes and incredulous stares.

2. Another cause of the relative paucity of actual *arguments* against the constructive empiricist's distinction between acceptance and belief is due to the fact that many commentators – a recent example is Blackburn (2002) – seem to understand the notion of acceptance as a kind of restricted belief (in this case, belief in the observable consequences of the theory, rather than belief in the theory *tout court*). This leads them to simply conflate the distinction between acceptance and belief with the distinction between observable and unobservable phenomena upon which – at least on their understanding – it must rest. Consequently, many arguments supposedly concerning the constructive empiricist's distinction between acceptance and belief are in fact arguments concerning the constructive empiricist's distinction between observable and unobservable phenomena, and thus quite irrelevant to our present concerns. Nevertheless, since any satisfactory response to Horwich will *ipso facto* serve to distinguish the two attitudes of acceptance and belief quite independently of the distinction between observables and unobservables, such confusion will be resolved *en passant* in the course of the main discussion.

3. Some philosophers, for example, Cohen (1992), have even argued that in fact *none* of the usual functional characterisations of belief are either necessary or sufficient for that mental state. In his view, belief is merely 'a disposition normally to feel that things are thus-and-so, not a disposition to say that they are or to act accordingly' (ibid.: 8). Although I agree in large part with Cohen's argument, I need not defend such a strong thesis here; my argument is merely that one cannot attribute *all* of the functional characterisations that those like Horwich have identified to our intuitive notion of belief.

4. Interestingly enough, such a line of thought is anticipated – and soundly rejected – in Horwich's original paper. Among the various attempts to distinguish between acceptance and belief which Horwich considers is the thought that, since we understand something like the constructive empiricist's

notion of acceptance in the context of scientific practice (employing an idealisation known to be false in order to facilitate easier calculation, for example), we can in fact understand the constructive empiricist's general notion of acceptance by generalising upon these specific and highly localised examples (1991: 4). Horwich, however, rejects this strategy: he argues that the ubiquity of the constructive empiricist's notion far outstrips anything in scientific practice that we could reasonably generalise upon; and, moreover, he also argues that the sort of picture one gets by extrapolating from these instances would not be constructive empiricism but a 'highly *radical* form of instrumentalism' where 'the acceptance of theories would not require the acceptance of *all* their logical consequences' and in which a unified understanding of nature is rejected in favour of 'the piecemeal understanding provided by a set of mini-theories that cannot be conjoined' (ibid.: 5). In Horwich's opinion then, the constructive empiricist should resist the sort of reasoning advanced by Teller, since this would result in a position *even worse* than before.

5. A similar point is made by van Dyck (2007: 25–26).
6. See, for example, Gilbert (2002) for a critical survey of the debate over acceptance and belief in collective intentionality. It is important to note that while Gilbert herself rejects the claim that groups are to be attributed the attitude of acceptance, at no point does she question that there is such an attitude to be contrasted with that of belief.
7. A similar argument to Bennett's was first put forward by Winters (1979). For other attempts to specify the precise manner in which belief is involuntary, see Scott-Kakures (1993), Raz (1999: 5–21) and Wedgwood (2002).
8. Lipton (2007: 118–119) appears to understand acceptance as a kind of restricted belief, with all of the associated difficulties that have been discussed above. Nevertheless, while I must disagree with Lipton insofar as I take acceptance and belief to be entirely distinct attitudes, the range of cases for which Lipton identifies the need for acceptance (on his understanding of the attitude) are readily appropriated with respect to our present understanding of the debate.
9. Detailed expositions can be found in Priest (2001: 139–161).
10. Such knowledge, arguably, will be conditional knowledge of similarity – knowledge of which combinatorially generated worlds, if indeed there are any, will count as relevantly similar to the actual world. For more, see Divers (2004: 672–673).
11. Let A be the proposition that we travel into outer space, and C be the proposition that we observe the moons. Then the first counterfactual, of the form $(A \,\square\!\!\rightarrow C)$, is analysed as the negative existential claim that there is no relevantly similar world where A is true and C is false, that is, $\neg \exists w(A \,\&\, \neg C)$. The second counterfactual is thus of the form $(A \,\square\!\!\rightarrow \neg C)$, which is then analysed as the negative existential claim that there is no world where A is true and C is true, that is, $\neg \exists w(A \,\&\, C)$. To *deny* the second counterfactual is therefore to assert the existence of a world where A and C are both true.
12. However, for a programmatic survey of some of the various strategies open to the uncommitted modal agnostic for dealing with the sort of modal statements that he cannot know the truth-values of, see Divers (2004: 675–683).

13. This is not to say of course that the committed modal agnostic is thereby committed to *every* modal statement, nor that the distinction between those modal statements that he is committed to and those that he is not is entirely arbitrary. The modal statements that he is committed to will be those that are consequences of the scientific theories that he accepts; thus, just as the constructive empiricist is only committed to certain statements about actual unobservable phenomena, on the grounds that they are consequences of those scientific theories that he believes to be empirically adequate, so too is the committed modal agnostic only committed to certain modal statements, on the grounds that they (and only they) are consequences of those theories.

14. For some technical issues in the exact formulation of modal fictionalism, see Brock (1993) and Rosen (1993); Hale (1995) raises an interesting dilemma over the modal status of the fictionalist's story about other possible worlds.

15. Although he does not dispute this response *per se*, Gibbard (1996: 333–334) does question the extent to which the modal quasi-realist can offer such a response and still maintain his anti-realism. For unlike the better-known programme of moral quasi-realism, the modal quasi-realist seems unable to provide a suitable anti-realist *explanation* of our modal discourse in non-modal terms. Gibbard's conclusion is that modal quasi-realism is therefore better construed as a kind of 'sophisticated realism', committed to uncovering the structure of our thought rather than providing an analysis of it. Since my aim in this section is merely to argue for the conceptual and/or semantic superiority of committed modal agnosticism over its anti-realist competitors, I shall not pursue this issue here.

16. Actually, there is a small caveat here. As Divers and Melia (2002) have shown, even the genuine modal realist cannot provide a *completely* non-circular analysis of our modal discourse. The basic problem is that if we are to allow the possibility of a denumerably infinite number of alien properties – which indeed seems reasonable – the genuine realist will lack the expressive resources to capture all of these possibilities without invoking some modal notions. It is unclear to me what the consequences of this result are.

17. For a detailed discussion of the conceptual resources of both genuine and actualist realism, see Divers (2002: 47–50, 181–195).

18. It should be noted that Divers (1999) does in fact provide a method for securing extensional reasoning within the scope of a fictionalist operator, but only by employing a primitive necessity operator. Such an approach is thus unavailable to the strong modal fictionalist, who attempts to analyse our modal discourse in terms of a possible worlds fiction, as Divers himself notes.

19. Although for some technical objections, see Hale (1993).

References

Adams, R. M. (1974) 'Theories of Actuality', *Noûs* **8**, pp. 211–231.

Anscombe, G. E. M. (1957) *Intention* (Oxford: Blackwell).

Balaguer, M. (1998) *Platonism and Anti-Platonism in Mathematics* (New York: Oxford University Press).

Benacerraf, P. (1973) 'Mathematical Truth', *Journal of Philosophy* **70**, pp. 661–679.

Bennett, J. (1990) 'Why Is Belief Involuntary?', *Analysis* **50**, pp. 87–107.

Blackburn, S. (1984) *Spreading the Word* (Oxford: Clarendon).

Blackburn, S. (1987) 'Morals and Modals', in G. MacDonald and C. Wright (eds) *Fact, Science and Morality: Essays on A. J. Ayer's Language, Truth and Logic* (Oxford: Blackwell), pp. 119–142.

Blackburn, S. (1988) 'Attitudes and Contents', *Ethics* **98**, pp. 501–517.

Blackburn, S. (1993) *Essays in Quasi-Realism* (New York: Oxford University Press).

Blackburn, S. (2002) 'Realism: Deconstructing the Debate', *Ratio* **15**, pp. 111–133.

Bloor, D. (1976) *Knowledge and Social Imagery* (London: Routledge).

Brock, S. (1993) 'Modal Fictionalism: A Reply to Rosen', *Mind* **102**, pp. 147–150.

Bueno, O. (1997) 'Empirical Adequacy: A Partial Structure Approach', *Studies in History and Philosophy of Science* **28**, pp. 585–610.

Bueno, O. (1999) 'Empiricism, Conservativeness and Quasi-Truth', *Philosophy of Science* **66**, pp. S474–S485.

Bueno, O. (2003) 'Is It Possible to Nominalise Quantum Mechanics?', *Philosophy of Science* **70**, pp. 1424–1436.

Carnap, R. (1936) 'Testability and Meaning', *Philosophy of Science* **3**, pp. 419–471.

Carnap, R. (1937) 'Testability and Meaning – Continued', *Philosophy of Science* **4**, pp. 1–40.

Cartwright, N. (1983) *How the Laws of Physics Lie* (Oxford: Clarendon Press).

Cartwright, N. (1999) *The Dappled World: A Study of the Boundaries of Science* (Cambridge: Cambridge University Press).

Chakravartty, A. (2004) 'Stance Relativism: Empiricism Versus Metaphysics', *Studies in History and Philosophy of Science* **35**, pp. 173–184.

Chakravartty, A. (2007) 'Six Degrees of Speculation: Metaphysics in Empirical Contexts', in B. Monton (ed.) *Images of Empiricism: Essays on Science and Stance, with a Reply from Bas C. van Fraassen* (Oxford: Oxford University Press), pp. 183–208.

Chalmers, D. (2002) 'Does Conceivability Entail Possibility?', in T. S. Gendler and J. Hawthorne (eds) *Conceivability and Possibility* (Oxford: Oxford University Press), pp. 145–200.

Chihara, C. and Chihara, C. (1993) 'A Biological Objection to Constructive Empiricism', *The British Journal for the Philosophy of Science* **44**, pp. 653–658.

Chisholm, R. M. (1946) 'The Contrary-to-Fact Conditional', *Mind* **55**, pp. 289–307.

Christensen, D. (1991) 'Clever Bookies and Coherent Beliefs', *Philosophical Review* **100**, pp. 229–247.

Churchland, P. M. (1985) 'The Ontological Status of Observables: In Praise of the Superempirical Virtues', in P. M. Churchland and C. A. Hooker (eds) *Images of Science: Essays on Realism and Empiricism with a Reply from Bas C. van Fraassen* (Chicago: University of Chicago Press), pp. 35–47.

Cohen, L. J. (1992) *An Essay on Belief and Acceptance* (Oxford: Clarendon Press).

da Costa, N. C. A. (1986) 'Pragmatic Probability', *Erkenntnis* 25, pp. 141–162.

da Costa, N. C. A. and French, S. (1989) 'Pragmatic Truth and the Logic of Induction', *The British Journal for the Philosophy of Science* 40, pp. 333–356.

da Costa, N. C. A. and French, S. (1990) 'The Model-Theoretic Approach in the Philosophy of Science', *Philosophy of Science* 57, pp. 248–265.

Dennett, D. (1978) *Brainstorms* (Cambridge, MA: MIT Press).

Dicken, P. (2008) 'Conditions May Apply: Essay Review of J. Ladyman & D. Ross *Every Thing Must Go*', *Studies in History and Philosophy of Science* 39, pp. 290–293.

Dicken, P. and Lipton, P. (2006) 'What Can Bas Believe? Musgrave and van Fraassen on Observability', *Analysis* 66, pp. 226–233.

Divers, J. (1995) 'Modal Fictionalism Cannot Deliver Possible Worlds Semantics', *Analysis* 55, pp. 81–88.

Divers, J. (1999) 'A Modal Fictionalist Result', *Noûs* 33, pp. 317–346.

Divers, J. (2002) *Possible Worlds* (London: Routledge).

Divers, J. (2004) 'Agnosticism About Other Worlds: A New Anti-Realist Programme in Modality', *Philosophy and Phenomenological Research* 49, pp. 659–684.

Divers, J. (2006) 'Possible Worlds Semantics Without Possible Worlds: The Agnostic Approach', *Mind* 115, pp. 187–225.

Divers, J. and Melia, J. (2002) 'The Analytic Limit of Genuine Modal Realism', *Mind* 111, pp. 15–36.

Eddington, A. (1928) *The Nature of the Physical World* (Ann Arbor: University of Michigan Press).

Einstein, A. (1916) 'Strahlungs-Emission und -Absorption nach der Quantentheorie', *Verhandlungen der deutschen Gesellschaften* 18, pp. 318–323.

Einstein, A. (1917) 'Zur Quantentheorie der Strahlung', *Physikalische Zeitschrift* 18, pp. 121–128.

Engel, P. (1997) 'P But I Shall Believe That Not P', Colloque Annuel de l'Institut Universitaire de France, Nantes (unpublished typescript).

Field, H. (1980) *Science Without Numbers: A Defence of Nominalism* (Oxford: Blackwell).

Field, H. (1982) 'Realism and Anti-Realism About Mathematics', *Philosophical Topics* 13, pp. 45–69.

Field, H. (1984a) 'Is Mathematical Knowledge Just Logical Knowledge?', *Philosophical Review* 93, pp. 509–522.

Field, H. (1984b) 'Gottlieb's *Ontological Economy: Substitutional Quantification and Mathematics*', *Noûs* 18, pp. 160–166.

Field, H. (1989) *Realism, Mathematics and Modality* (Oxford: Blackwell).

Fine, K. (1977) 'Prior on the Construction of Possible Worlds and Instants', in A. N. Prior and K. Fine (eds) *Worlds, Times and Selves* (London: Duckworth), pp. 116–161.

Fodor, J. A. (1983) *The Modularity of Mind* (Cambridge, MA: MIT Press).

Forbes, G. (1989) *Languages of Possibility* (Oxford: Blackwell).

Garcia-Carpintero, M. and Macia, J. (2004) *Two-Dimensional Semantics* (Oxford: Oxford University Press).

Geach, P. (1960) 'Ascriptivism', *Philosophical Review* 69, pp. 221–225.

Geach, P. (1965) 'Assertion', *Philosophical Review* 74, pp. 449–465.

Gibbard, A. (1996) 'Projectivism, Quasi-Realism and Sophisticated Realism', *Mind* 105, pp. 331–335.

Giere, R. N. (1985) 'Constructive Realism', in P. M. Churchland and C. A. Hooker (eds) *Images of Science: Essays on Realism and Empiricism with a Reply from Bas C. van Fraassen* (Chicago: University of Chicago Press), pp. 75–98.

Giere, R. N. (1988) *Explaining Science: A Cognitive Approach* (Chicago: University of Chicago Press).

Giere, R. N. (1999) *Science Without Laws* (Chicago: University of Chicago Press).

Gilbert, M. (2002) 'Belief and Acceptance as Features of Groups', *Protosociology* 16, pp. 35–69.

Goldstein, M. (1983) 'The Prevision of a Prevision', *Journal of the American Statistical Association* 78, pp. 817–819.

Goodman, N. (1947) 'The Problem of Counterfactual Conditionals', *Journal of Philosophy* 44, pp. 113–120; reprinted in his (1954) *Fact, Fiction and Forecast* (Cambridge, MA: Harvard University Press), pp. 3–27.

Goodman, N. (1954) *Fact, Fiction and Forecast* (Cambridge, MA: Harvard University Press).

Gottlieb, D. (1980) *Ontological Economy: Substitutional Quantification and Mathematics* (Oxford: Clarendon).

Hacking, I. (1985) 'Do We See Through a Microscope?', in P. M. Churchland and C. A. Hooker (eds) *Images of Science: Essays on Realism and Empiricism with a Reply from Bas C. van Fraassen* (Chicago: University of Chicago Press), pp. 132–152.

Hale, B. (1993) 'Can There Be a Logic of Attitudes?', in J. Haldane and C. Wright (eds) *Reality, Representation and Projection* (Oxford: Oxford University Press), pp. 337–363.

Hale, B. (1995) 'Modal Fictionalism: A Simple Dilemma', *Analysis* 55, pp. 63–67.

Hawthorne, J. (1996) 'The Epistemology of Possible Worlds: A Guided Tour', *Philosophical Studies* 84, pp. 183–202.

Hazen, A. (1976) 'Expressive Completeness in Modal Languages', *Journal of Philosophical Logic* 5, pp. 25–46.

Hempel, C. (1958) 'The Theoretician's Dilemma: A Study in the Logic of Theory Construction', in H. Feigl, M. Scriven and G. Maxwell (eds) *Concepts, Theories and the Mind-Body Problem*, Minnesota Studies in the Philosophy of Science Vol. II (Minneapolis: University of Minnesota Press), pp. 37–99.

Hieronymi, P. (2006) 'Controlling Attitudes', *Pacific Philosophical Quarterly* 87, pp. 45–74.

Hintikka, J. (1969) *Models for Modalities* (Dordrecht: D. Reidel Publishing).

Ho, D. (2007) 'Farewell to Empiricism', in B. Monton (ed.) *Images of Empiricism: Essays on Science and Stance, with a Reply from Bas C. van Fraassen* (Oxford: Oxford University Press), pp. 319–333.

Horwich, P. (1991) 'On the Nature and Norms of Theoretical Commitment', *Philosophy of Science* 58, pp. 1–14; reprinted in his (2005) *From a Deflationary Point of View* (Oxford: Oxford University Press), pp. 86–104.

Howson, C. (2000) *Hume's Problem: Induction and the Justification of Belief* (Oxford: Oxford University Press).

Jackson, F. (1986) 'What Mary Didn't Know', *Journal of Philosophy* **83**, pp. 291–295.

James, W. (1948) 'The Will to Believe', in his *Essays in Pragmatism* (New York: Hafner Press), pp. 88–109; originally published in *New World* 5, pp. 327–347.

Jammer, M. (1966) *The Conceptual Development of Quantum Mechanics* (New York: McGraw-Hill).

Jauernig, A. (2007) 'Must Empiricism Be a Stance, and Could it Be One? How to Be an Empiricist and a Philosopher at the Same Time', in B. Monton (ed.) *Images of Empiricism: Essays on Science and Stance, with a Reply from Bas C. van Fraassen* (Oxford: Oxford University Press), pp. 271–318.

Kahneman, D., Slovic, P. and Tversky, A. (eds) (1982) *Judgement Under Uncertainty: Heuristics and Biases* (Cambridge: Cambridge University Press).

Kreisel, G. (1967) 'Informal Rigour and Completeness Proofs', in I. Lakatos (ed.) *Problems in the Philosophy of Mathematics* (Amsterdam: North Holland), pp. 138–171.

Kuhn, T. (1962) *The Structure of Scientific Revolutions* (Chicago: University of Chicago Press).

Kuhn, T. (1978) *Black-Body Theory and the Quantum Discontinuity 1894–1912* (Oxford: Oxford University Press).

Kukla, A. (1998) *Studies in Scientific Realism* (Oxford: Oxford University Press).

Ladyman, J. (2000) 'What's Really Wrong with Constructive Empiricism? Van Fraassen and the Metaphysics of Modality', *The British Journal for the Philosophy of Science* **51**, pp. 837–856.

Ladyman, J. (2004a) 'Constructive Empiricism and Modal Metaphysics: A Reply to Monton and van Fraassen', *The British Journal for the Philosophy of Science* **55**, pp. 755–765.

Ladyman, J. (2004b) 'Empiricism Versus Metaphysics', *Philosophical Studies* **121**, pp. 133–145.

Ladyman, J. (2007) 'The Epistemology of Constructive Empiricism', in B. Monton (ed.) *Images of Science: Essays on Science and Stance, with a Reply from Bas C. van Fraassen* (Oxford: Oxford University Press), pp. 46–61.

Ladyman, J., Douven, I., Horsten, L. and van Fraassen, B. C. (1997) 'A Defence of van Fraassen's Critique of Abductive Reasoning: Reply to Psillos', *Philosophical Quarterly* **47**, pp. 305–321.

Ladyman, J. and Ross, D. (2007) *Every Thing Must Go: Metaphysics Naturalised* (Oxford: Oxford University Press).

Levi, I. (1987) 'The Demons of Decision', *The Monist* **70**, pp. 193–211.

Lewis, D. (1978) 'Truth in Fiction', *American Philosophical Quarterly* **15**, pp. 37–46; reprinted in his (1983) *Philosophical Papers, Vol. I* (Oxford: Oxford University Press), pp. 261–280.

Lewis, D. (1980) 'Mad Pain and Martian Pain', in N. Block (ed.) *Readings in Philosophy of Psychology, Vol. I* (Cambridge, MA: Harvard University Press), pp. 216–222.

Lewis, D. (1986) *On the Plurality of Worlds* (Oxford: Blackwell).

Lipton, P. (1993) 'Is the Best Good Enough?', *Proceedings of the Aristotelian Society* **93**, pp. 89–104.

Lipton, P. (2000) 'Tracking Track Records', *Proceedings of the Aristotelian Society* **74** (supplementary volume), pp. 179–205.

Lipton, P. (2004) *Inference to the Best Explanation* (London: Routledge).

Lipton, P. (2007a) 'Accepting Contradictions', in B. Monton (ed.) *Images of Empiricism: Essays on Science and Stances, with a Reply from Bas C. van Fraassen* (Oxford: Oxford University Press), pp. 117–133.

Lipton, P. (2007b) 'Science and Religion: The Immersion Solution', in A. Moore and M. Scott (eds) *Realism and Religion: Philosophical and Theological Perspectives* (Aldershot: Ashgate), pp. 31–46.

Mach, E. (1893) *The Science of Mechanics* (La Salle: Open Court).

Mach, E. (1910) *Popular Scientific Lectures* (Chicago: Open Court).

Mackie, J. L. (1973) *Truth, Probability and Paradox* (Oxford: Clarendon Press).

Maher, P. (1992) 'Diachronic Rationality', *Philosophy of Science* **59**, pp. 120–141.

Malament, D. (1982) 'Review of Field's *Science Without Numbers*', *Journal of Philosophy* **79**, pp. 523–534.

Maxwell, G. (1962) 'The Ontological Status of Theoretical Entities', in H. Feigl and G. Maxwell (eds) *Scientific Explanation, Space and Time, Minnesota Studies in the Philosophy of Science Vol. III* (Minneapolis: University of Minnesota Press), pp. 3–27.

Melchert, N. (1985) 'Why Constructive Empiricism Collapses into Scientific Realism', *Australasian Journal of Philosophy* **63**, pp. 213–215.

Melia, J. (1992) 'Against Modalism', *Philosophical Studies* **68**, pp. 35–56.

Melia, J. (1998) 'Field's Programme: Some Interference', *Analysis* **58**, pp. 63–71.

Melia, J. (2003) *Modality* (Chesham: Acumen).

Mellor, D. H. (1988) 'The Warrant of Induction', Inaugural Lecture, University of Cambridge; reprinted in his (1991) *Matters of Metaphysics* (Cambridge: Cambridge University Press), pp. 254–268.

Misak, C. (2004) *Truth and the End of Inquiry: A Peircean Account of Truth* (Oxford: Oxford University Press).

Mohler, C. (2007) 'The Dilemma of Empiricist Belief', in B. Monton (ed.) *Images of Empiricism: Essays on Science and Stances, with a Reply from Bas C. van Fraassen* (Oxford: Oxford University Press), pp. 209–228.

Monton, B. and van Fraassen, B. C. (2003) 'Constructive Empiricism and Modal Nominalism', *The British Journal for the Philosophy of Science* **54**, pp. 405–422.

Muller, F. A. (2004) 'Can a Constructive Empiricist Adopt the Concept of Observability?', *Philosophy of Science* **71**, pp. 80–97.

Muller, F. A. (2005) 'The Deep Black Sea: Observability and Modality Afloat', *The British Journal for the Philosophy of Science* **56**, pp. 61–99.

Muller, F. A. and van Fraassen, B. C. (2008) 'How to Talk About Unobservables', *Analysis* **68**, pp. 197–205.

Musgrave, A. (1985) 'Realism Versus Constructive Empiricism', in P. M. Churchland and C. A. Hooker (eds) *Images of Science: Essays on Realism and Empiricism with a Reply from Bas C. van Fraassen* (Chicago: Chicago University Press), pp. 197–221.

Nagel, E. (1950) 'Science and Semantic Realism', *Philosophy of Science* **17**, pp. 174–181.

Nagel, J. (2000) 'The Empiricist Conception of Experience', *Philosophy* **75**, pp. 345–376.

Nolan, D. (1997) 'Three Problems for "Strong" Modal Fictionalism', *Philosophical Studies* **87**, pp. 259–275.

Nolan, D. (2002) *Topics in the Philosophy of Possible Worlds* (New York: Routledge).

Peacocke, C. (1978) 'Necessity and Truth Theories', *Journal of Philosophical Logic* **7**, pp. 473–500.

Peirce, C. S. (1934) 'What Pragmatism Is', in C. Hartshorne and P. Weiss (eds) *Collected Papers of Charles Sanders Peirce, Vol. V: Pragmatism and Pragmaticism* (Cambridge, MA: Harvard University Press), pp. 272–292; originally published in *The Monist* **15**, pp. 161–181.

Planck, M. (1900a) 'Zur Theorie des Gesetzes der Energieverteilung im Normalspektrum', *Verhandlungen der deutschen Physikalischen Gesellschaft* **2**, pp. 237–245.

Planck, M. (1900b) 'Uber irreversible Strahlungsvorgange', *Annalen der Physik* **1**, pp. 69–122.

Planck, M. (1901) 'Vereinfachte Ableitung der Schwingungsgesetze eines linearen Resonators im stationaren Feld', *Physikalische Zeitschrift* **2**, pp. 530–534.

Plantinga, A. (1974) *The Nature of Necessity* (Oxford: Clarendon).

Plantinga, A. (1993) *Warrant: The Current Debate* (Oxford: Oxford University Press).

Popper, K. (1959) *The Logic of Scientific Discovery* (London: Hutchinson).

Priest, G. (2001) *An Introduction to Non-Classical Logic* (Cambridge: Cambridge University Press).

Psillos, S. (1996) 'On van Fraassen's Critique of Abductive Reasoning', *The Philosophical Quarterly* **46**, pp. 31–47.

Psillos, S. (1999) *Scientific Realism: How Science Tracks Truth* (London: Routledge).

Psillos, S. (2007) 'Putting a Bridle on Irrationality: An Appraisal of van Fraassen's New Epistemology', in B. Monton (ed.) *Images of Empiricism: Essays on Science and Stances, with a Reply from Bas C. van Fraassen* (Oxford: Oxford University Press), pp. 134–164.

Putnam, H. (1963) 'Degrees of Confirmation and Inductive Logic', in P. A. Schilpp (ed.) *The Philosophy of Rudolph Carnap* (La Salle, IL: Open Court), pp. 761–783.

Raz, J. (1999) *Engaging Reasons* (Oxford: Oxford University Press).

Reichenbach, H. (1938) *Experience and Prediction* (Chicago: University of Chicago Press).

Reichenbach, H. (1949) *The Theory of Probability* (Berkeley: University of California Press).

Rosen, G. (1990) 'Modal Fictionalism', *Mind* **99**, pp. 327–354.

Rosen, G. (1993) 'A Problem for Fictionalism About Possible Worlds', *Analysis* **53**, pp. 71–81.

Rosen, G. (1994) 'What Is Constructive Empiricism?', *Philosophical Studies* **74**, pp. 143–178.

Rosen, G. (1998) 'Blackburn's *Essays in Quasi-Realism*', *Noûs* **32**, pp. 386–405.

Routley, R. (1980) *Exploring Meinong's Jungle: An Investigation of Noneism and the Theory of Items* (Canberra: Australian National University Press).

Scott-Kakures, D. (1993) 'On Belief and the Captivity of the Will', *Philosophy and Phenomenological Research* **53**, pp. 77–103.

Shapiro, S. (1983) 'Conservativeness and Incompleteness', *Journal of Philosophy* **80**, pp. 521–531.

Smith, J. (1988) 'Inconsistency and Scientific Reasoning', *Studies in History and Philosophy of Science* 19, pp. 429–445.

Stalnaker, R. C. (1972) 'Pragmatics', in G. Harman and D. Davidson (eds) *Semantics of Natural Language* (Dordrecht: Reidel), pp. 380–397.

Stalnaker, R. C. (1974) 'Pragmatic Presuppositions', in P. Unger and M. Munitz (eds) *Semantics and Philosophy* (New York: New York University Press), pp. 197–213.

Talbott, W. J. (1991) 'Two Principles of Bayesian Epistemology', *Philosophical Studies* 62, pp. 135–150.

Tarski, A. (1936) 'On the Concept of Logical Consequence', in his *Logic, Semantics, Metamathematics* (Indianapolis: Hackett Publishing), pp. 409–420.

Teller, P. (2001a) 'Whither Constructive Empiricism?', *Philosophical Studies* 106, pp. 123–150.

Teller, P. (2001b) 'Twilight of the Perfect Model Model', *Erkenntnis* 55, pp. 393–415.

van Cleve, J. (1984) 'Reliability, Justification and the Problem of Induction' in P. French, T. Uehling and H. Wettstein (eds) *Midwest Studies in Philosophy, Vol. IX* (Minneapolis: University of Minnesota Press), pp. 555–567.

van Dyck, M. (2007) 'Constructive Empiricism and the Argument from Underdetermination', in B. Monton (ed.) *Images of Empiricism: Essays on Science and Stances, with a Reply from Bas C. van Fraassen* (Oxford: Oxford University Press), pp. 11–31.

van Fraassen, B. C. (1974) 'Platonism's Pyrrhic Victory', in A. R. Anderson, R. B. Marcus and R. M. Martin (eds) *The Logical Enterprise* (New Haven: Yale University Press), pp. 39–50.

van Fraassen, B. C. (1977a) 'The Pragmatics of Explanation', *American Philosophical Quarterly* 14, pp. 143–150.

van Fraassen, B. C. (1977b) 'The Only Necessity Is Verbal Necessity', *Journal of Philosophy* 74, pp. 71–85.

van Fraassen, B. C. (1978) 'Essence and Existence', in N. Rescher (ed.) *Studies in Ontology* (Oxford: Blackwell), pp. 1–25.

van Fraassen, B. C. (1980) *The Scientific Image* (Oxford: Clarendon Press).

van Fraassen, B. C. (1981) 'Essences and the Laws of Nature', in R. Healey (ed.) *Reduction, Time and Reality: Studies in the Philosophy of the Natural Sciences* (Cambridge: Cambridge University Press), pp. 189–200.

van Fraassen, B. C. (1984) 'Belief and the Will', *Journal of Philosophy* 81, pp. 235–256.

van Fraassen, B. C. (1985) 'Empiricism in the Philosophy of Science', in P. M. Churchland and C. A. Hooker (eds) *Images of Science: Essays on Realism and Empiricism with a Reply from Bas C. van Fraassen* (Chicago: University of Chicago Press), pp. 245–308.

van Fraassen, B. C. (1989) *Laws and Symmetry* (Oxford: Clarendon Press).

van Fraassen, B. C. (1991) *Quantum Mechanics: An Empiricist View* (Oxford: Clarendon Press).

van Fraassen, B. C. (1994) 'Against Transcendental Empiricism' in T. J. Stapleton (ed.) *The Question of Hermeneutics* (Dordrecht: Kluwer), pp. 309–335.

van Fraassen, B. C. (1995a) 'Belief and the Problem of Ulysses and the Sirens', *Philosophical Studies* 77, pp. 7–37.

van Fraassen, B. C. (1995b) 'Against Naturalised Epistemology' in P. Leonardi and M. Santambrogio (eds) *On Quine* (Cambridge: Cambridge University Press), pp. 68–88.

van Fraassen, B. C. (2000) 'The False Hopes of Traditional Epistemology', *Philosophy and Phenomenological Research* **60**, pp. 253–280.

van Fraassen, B. C. (2002) *The Empirical Stance* (New Haven: Yale University Press).

van Fraassen, B. C (2004a) 'Replies to Discussion on *The Empirical Stance*', *Philosophical Studies* **121**, pp. 171–192.

van Fraassen, B. C. (2004b) 'Reply to Chakravartty, Jauernig and McMullin', Symposium on *The Empirical Stance*, Pacific APA, Pasadena (unpublished typescript).

van Fraassen, B. C. (2007) 'From a View of Science to a New Empiricism', in B. Monton (ed.) *Images of Empiricism: Essays on Science and Stances, with a Reply from Bas C. van Fraassen* (Oxford: Oxford University Press), pp. 337–383.

van Fraassen, B. C. (2008) *Scientific Representation: Paradoxes of Perspective* (Oxford: Oxford University Press).

Wedgwood, R. (2002) 'The Aim of Belief', *Philosophical Perspectives* **16**, pp. 267–297.

Williams, B. A. O. (1973) 'Deciding to Believe', in his *Problems of the Self* (Cambridge: Cambridge University Press), pp. 136–151.

Winters, B. (1979) 'Willing to Believe', *Journal of Philosophy* **76**, pp. 243–256.

Worrall, J. (1989) 'Structural Realism: The Best of Both Worlds?', *Dialectica* **43**, pp. 99–124.

Wright, C. (1988) 'Realism, Anti-Realism, Irrealism, Quasi-Realism', *Midwest Studies in Philosophy* **12**, pp. 25–49.

Yablo, S. (1993) 'Is Conceivability a Guide to Possibility?', *Philosophy and Phenomenological Research* **53**, pp. 1–42.

Index